Extended Tools

in *Surface Analysis*

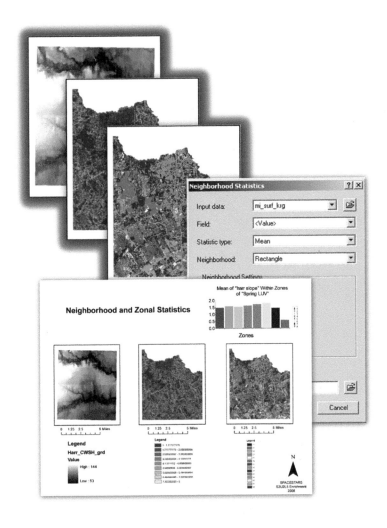

Hanebuth
Rotzler
Smith

Statement of Copyright

Extended Tools in Surface Analysis

Table of Contents - Student Manual

Applications in Surface Analysis

An in-depth look at surface and various raster data that focuses on common and useful tools essential for geospatial technicians. Each lesson features detailed narratives of tools in the context of key processes, an exercise applying the concepts, an enrichment lesson, and a lesson review to reinforce key concepts.

Glossary:

Preface

This book directs you through five types of applications in Surface Analysis using the ArcGIS Spatial Analyst software extension. Emphasis is placed on the various methods and uses of displaying continuous, or grid, data over a surface. You will be able to map data such as elevation, rainfall, and temperature – data that differs from one location to the next on the surface of the earth. The five types of analyses that you will be using are: mapping distance, density, interpolation, surface analysis, and statistics. This book will conclude with a short project where you will use the skills you have acquired to perform a mini project combining all skills.

Lessons are built around applications to show process beyond "buttonology". This type of learning promotes knowledge of tools as well as building a solid foundation for users to make good decisions when faced with choices that will ultimately affect the end users, customers, and colleagues who will benefit from your analysis. The skills that that are taught in this book allow the user to go directly from classroom application to real world application.

This book is designed for secondary, post secondary, and professionals. There are neither prerequisites nor is previous GIS experience necessary. We provide all directions and data needed to complete the directed tasks.

How will I use this book?
This book is designed to be completed sequentially with lessons that provide an overview of ArcGIS and the Spatial Analyst for ArcGIS with subsequent lessons examining introductory processes in Surface Analysis. Each lesson will focus on a necessary concept that is fundamental to a users growth as geospatial technician.

Features in this book:

- **Step-by-Step Instructions** –an easy to follow format relevant for novice to experienced ArcGIS users
- **Lesson Content** – provides a narrative overview of an important topic relevant to the fundamental concepts of geospatial technology
- **Lesson Exercise** – applies lesson content to an exercise that features "real world" tasks at work
- **Lesson Review** – reinforces knowledge gained with exercises to identify key terms and concepts
- **"Knowledge Knugget" boxes** – boxes found in the margin that features tools, tips and tricks that may enhance your experience with geospatial technologies
- **Finished Layouts** – to provide a self assessment tool to ensure successful completion of each lesson

About ESRI's ArcGIS

ArcGIS is a software suite developed by Environmental Services Research Institute, Inc. (ESRI) designed to analyze and model geospatial data. Of the software suite in ArcGIS, this book will use three major components; ArcMap, ArcToolbox and ArcCatalog. ArcMap is the primary part of the suite that will be used throughout the book to display, create, and analyze different types of geospatial data. ArcToolbox contains various geoprocessing tools used throughout the ArcGIS suite to complete various tasks such as creating buffers, merging shapefiles, and address locators. ArcCatalog is the "virtual filing cabinet" where users create, manipulate, or preview data and metadata. ArcGIS is widely accepted and used among today's GIS professionals and students. Using this software will make for a smooth transition for the student to take their GIS skills from the classroom to the workplace or other academic pursuits.

Teacher Materials

The STARS Introduction to Geographic Information Systems and Remote Sensing Concepts is designed to be used in a variety of learning environments. For classroom environments, a teacher's edition is available with the following enhancements:

Overviews – Each lesson in the teacher's edition comes with a lesson overview page for instructors with boxes in the margins that provide quick reference to lesson goals. The "What Will You Teach" section provides the instructor with a bulleted list that includes a list of goals for the lesson GIS skills learned. The "How Will You Teach It" section provides the instructor with the procedures to introduce the topic and have the students complete the lesson. The initial introduction provides a unique perspective on the topic that is covered in the lesson that will enhance teacher's abilities to facilitate learning in a classroom environment. The overview also describes in detail the skills taught in the lesson as well as additional information that may be necessary to complete the lesson. Lesson related links for further study are also supplied with each overview page.

PowerPoint Presentation Notes – Each student lesson comes with a PowerPoint presentation that provides an overview of the lesson including concepts, the skills they will cover, and the study area involved. The teacher's manual is supplemented with detailed descriptions and commentary for each slide allowing a diverse range of instructors to lead classroom lecture. For every presentation, the student manuals will have notes sheets, complete with thumbnail pictures of the slides from the presentation, with lines for notes beside them.

Assessments – Lessons will conclude with a full page color layout of a successfully completed exercise. If questions are presented within a lesson, the teacher's manual includes answers to those questions.

About the Authors

Eddie Hanebuth is founder and president of Digital Quest, a Mississippi-based development and training-oriented company that produces GIS instructional material for educational institutions. He chairs the U.S. Department of Labor's National Standard Geospatial Apprenticeship Program and the SkillsUSA Geospatial Competition Committee, and runs the SPACESTARS teacher-training laboratory in the Center of Geospatial Excellence, NASA's John C. Stennis Space Center.

Liz Rotzler has six years experience in geospatial technology education. After teaching GIS in the classroom using the STARS curriculum and certification, she has spent the past three years working in development of GIS/RS curriculum. She has co-authored and edited the first book in Digital Quest's aGIS series, Introduction to Geospatial Technologies as well as worked in the popular Digital Quest SPACESTARS series.

Austin Smith has been part of the Digital Quest team for four years. He is currently the Vice President of Development and Support and also serves as the chair of the S.T.A.R.S. Geospatial Certification Committee. He has experience in information technology development, implementation, and training in a variety of public and private organizations. At Digital Quest he has co-authored or edited over 30 titles. With Digital Quest's STARS series, he serves in authoring, planning, and final editing.

Acknowledgements

Digital Quest wishes to express their appreciation to Alison Hayes of Geographic Magic for her work in editing this book.

Applications in Surface Analysis

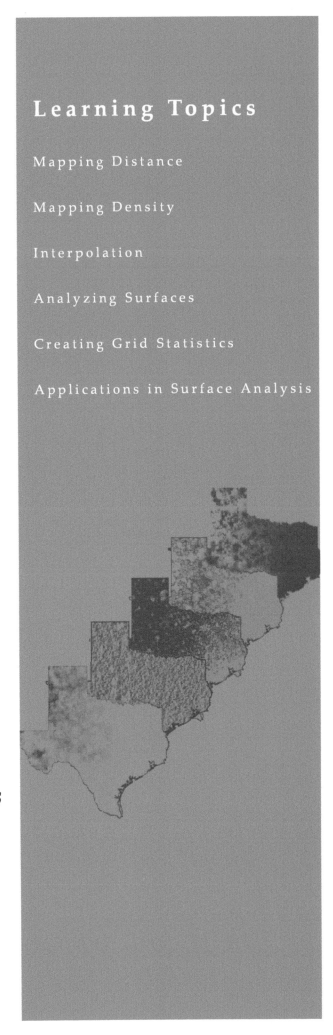

Extended Tools in Surface Analysis

Lesson 1:
Mapping Distance

The Spatial Analyst extension is used to take into account distances and to calculate the relationship between locations. You know that while drawing on paper, the shortest distance between two points is a straight line. However, the real world does not always function like that, and spatial analysis software helps you to take this into account. For example, your car is best served by remaining on a road, and while the supermarket might be directly east of you, you're not going to drive due east if that involves things besides roads, such as your neighbors, other houses, swimming pools, and possibly a stream or cliff. When you are determining distances using Spatial Analyst, you can take into account the relief of the surface of the study area. Using elevation grids allows you to get a more accurate measurement based on a model of what the surface of the Earth actually looks like for the study area. Distance in Spatial Analyst can be measured as straight line distance or cost weighted distance. You will explore both of these types of analysis in this lesson.

The straight line distance function is also called "Euclidean Distance," which gives the distance from each cell in the raster grid to the closest point of interest. This is also referred to distance 'as the crow flies' – which boils down to an ideal world where everything is flat and you can simply go from point A to point B. As this does not normally happen, we need to take the obstacles into account when making a realistic distance-related plan. This is known as the cost-weighted distance. Cost-weighted distance is particularly helpful when you want to locate something in an area that has the least travel cost to a particular location. The graphic is displaying Cost Allocation for each cell which source (hospitals in this case) is the least costly to reach based on the cost of the slope.

While they are similar, the Cost Weighted grid surface shows greater access variation based on cost constraints that were not considered in the straight line analysis. Distance can be calculated based on other cost factors such as slope, terrain, traffic lights, etc.

Cost weighted distance is relevant when determining the cost distance to this site from a single potential assisted living facility in the same cost allocation zone.

Shortest path analysis is particularly helpful for applications such as finding the best location and route for a new road.

Lesson 1: Mapping Distance

In earlier lessons dealing with ArcGIS 9x software programs, you have used the Measure Tool with feature data (shapefiles) to determine distances between features on a map. The measurements that were taken were estimates, of course, as these measurements were based on measuring on a flat surface and did not take into consideration that the Earth has certain bumps and indentations that will vary the distance somewhat. When determining distances using Spatial Analyst, you can take into account the relief of the surface of the study area. Using elevation grids allows you to get a more accurate measurement based on a model of what the surface of the Earth actually looks like for the study area. Distance in Spatial Analyst can be measured as straight line distance or cost weighted distance. You will explore both of these types of analysis in this lesson.

Loading Spatial Analyst Extension

Before any type of analysis using Spatial Analyst can begin, you must load the Spatial Analyst extension program. To do this,

1.   *Launch* **ArcMap** with a new empty map.

2.   *Save* the project as **S3SAL1_XX.mxd** (where XX is your initials) in your **student folder**.

3.   *Rename* this data frame **Straight Line Distance**. *Click* [OK] to close.

4.   *Select* **Extensions...** from the **Tools** menu.

5.   *Click* the **checkbox** to the left of **Spatial Analyst** to select this extension.

6.   *Click* [Close].

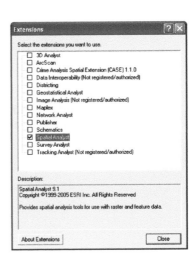

To view the **Spatial Analyst toolbar** in the ArcMap window,

7.   *Select* **Toolbars/Spatial Analyst** from the **View** menu.

8.   *Click* [Close].

Straight Line Distance

Straight line distance calculations create buffers around a source layer to identify certain distance ranges. Straight line distance calculations do not take into consideration any "costs" that may be associated with traveling from one point to another over a surface. This type of distance function is also called "Euclidean Distance," which gives the distance from each cell in the raster grid to the closest point of interest. It can be thought of the distance from one place to another "as the crow flies."

Before you begin any analysis, you should specify the location of the working directory that you would like to use for the temporary files that will be created during some of the analyses. To do this,

1. **Select** **Options...** from the  drop-down menu.

2. **Click** the **General** tab.

   In the **Working directory** box,

3. **Enter** or **browse** to specify your **student folder**.

4. **Click** **OK** .

5. **Add** the **laud_dem** grid layer and the **Lauderdale_Hospitals.shp** layer to the map from the **C:\STARS\ExtToolsinSA** folder.

6.   To determine the cell size of the **laud_dem** grid,
      double click the layer in the Table of Contents to open **Layer Properties**.

      Click the **Source** tab.

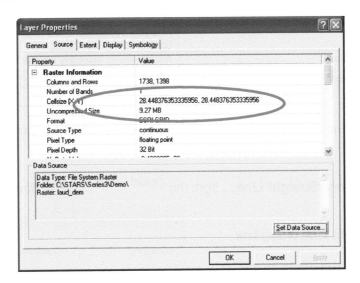

7.   ***Click*** [OK] to close the window.

Notice that the cell size of this grid is ***28.448376353335998*** we will round it to ***28.448376***. Cell size refers to the resolution of a grid or, in map units, how much one cell in the grid represents in ground area. In conducting analysis with grid data using Spatial Analyst, it is important to remember that you should always use the largest cell size of the input rasters in determining the cell size of output layers. This is because all raster grids will resample up to the greatest cell size when combined for analysis. In this case, only one (1) grid (raster) is being used; therefore, the output cell size should be set to the cell size of this grid when performing analysis.

In order to determine distances to hospitals for the entire study area (in this case, all of Lauderdale County, MS), you must set the Analysis Extent prior to performing the Distance analysis.

To set the extent of the analysis,

8.   ***Select*** **Options...** from the Spatial Analyst ▼ drop-down menu.

9.   ***Click*** the **Extent** tab.

10.   From the **Analysis extent** drop-down list, ***Select*** **Same as Layer "laud_dem."**

11.   ***Click*** [OK] to close the window.

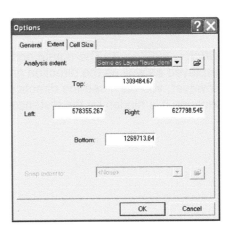

You set the Analysis Extent to the extent of the grid layer that provides elevation data for the entire county of Lauderdale, MS. Other options for setting Analysis Extent include:

Analysis Extent Option	Output
Intersection of Inputs	Analysis performed only to the extent where all of the layers included overlay
Union of Inputs	Analysis performed to the extent of all layers combined
As Specified Below	Allows you to manually set the bounding coordinates where analysis should take place
Same as Layer "Filename"	Analysis performed to the extent of one of the layers

You will now determine the straight line distances to each hospital from locations throughout the county.

12. **Select Distance/Straight Line...** from the Spatial Analyst ▼ drop-down menu.

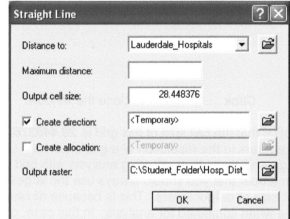

13. In the **Straight Line** dialog box, **specify Lauderdale_Hospitals** as the **Distance to** layer.

14. **Enter 28.448376** as the **output cell size**

15. **Click** to **Create direction** but leave the output raster for direction as **temporary**.

16. **Save** the **output raster** as **Hosp_Dist_SL** in your **student folder**.

Direction creates a grid that illustrates the direction from each cell to the closest source feature (in this case, the closest hospital).

17. **Click** OK . The output grid displaying distance to the nearest hospital is displayed as well as the direction grid. However, you can't see the direction grid because it is hidden by the distance grid.

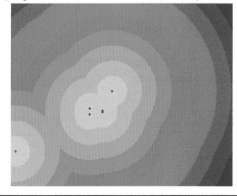

By looking at the bottom right portion of the ArcMap window you will notice that the display units for this map are meters (because DEM data is projected in UTM coordinates that are expressed in meters). Therefore the graduated classifications provided in the Table of Contents for distance to the closest hospital are expressed in meters.

In this scenario, this new layer displays the distance in bands extending approximately 3,300 meters each from every hospital location in the county. Most accurately, however, what is truly represented here is each cell's distance in meters from the closest hospital. The values of each cell are not uniform, as would be the case if this were a vector polygon. Since this is a grid surface, each cell has a unique value based on it's distance from the closest hospital. The color bands are used only to classify to the eye how these distances are calculated.

18.   ***Use*** the **Identify Tool** 🛈 to click around the grid layer to see that each cell has a unique distance value, even two cells in the same color band.

19.   In the Table of Contents ***turn off*** the **Dist_Hosp_SL** layer so that you can see the direction grid.

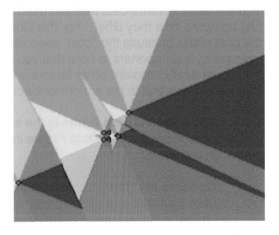

The legend under the direction layer name provides information on how to read the direction grid. Each color associates the direction from each cell to the closest hospital. The colors are classified on the basis of compass direction in degrees *(eg. South 157.5-202.5)*, with each cell having a unique direction value within the range of each direction class.

To make the direction grid permanent (remember that you left the output grid as temporary),

20.   ***Right click*** this layer (**Direction to Lauderdale_Hospital**) in the Table of Contents

21.   ***Select Data>*****Make Permanent** in the menu.

22.   ***Navigate*** to your **student folder**

23.   ***Name*** the grid **Hosp_Dir_SL**.

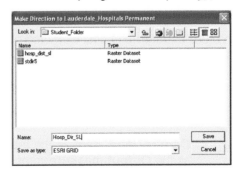

24.   ***Click*** Save .

Note: The Hosp_Dist_Sl data layer is already permanent.

Cost Weighted Distance & Allocation

Cost weighted distance "calculates a value for every cell that is the least accumulated cost of traveling from each cell to the source." (*Using ArcGIS Spatial Analyst*, 130). This is particularly helpful when you want to locate something in an area that has the least travel cost to a particular location.

For example, in conducting economic development studies you may identify several potential sites for locating a new business. When a business finds a new site, there is potential that suppliers and other support businesses will locate near the facility. Depending on the type of manufacturer and the type of suppliers involved, these businesses have different strategies concerning how far they typically locate to a manufacturer. This type of analysis can help with this.

You will use the same data that you used for the Straight Line Distance analysis for this analysis and compare how they differ. For this Cost Weighted Distance calculation, you will use slope as the cost raster because the "cost" associated with traversing a surface increases as slope increases. It is important to note that you will use this one (1) factor for simplicity purposes. You could easily consider other factors and datasets in compiling a cost raster for this analysis. You will explore that in the enrichment activity.

1. ***Add*** a second **data frame** to the in the project by selecting ***Data Frame*** in the ***Insert menu.***

2. ***Collapse*** the ***Straight Line Distance Data frame*** so you can see the new Data Frame.

3. ***Rename*** this second data frame **Cost Weighted Distance**.

4. ***Add*** the **laud_dem** and **Lauderdale_Hospitals.shp** layers to the data frame.

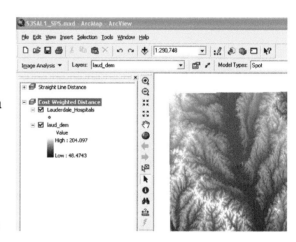

In order to determine cost weighted distance based on the terrain of Lauderdale County, the slope of the terrain (based on the elevation grid) must be determined.

5. ***Select* Surface Analysis/Slope…** from the Spatial Analyst ▼ drop-down menu.

6. ***Confirm* laud_dem** as the **Input surface** and **Degree** as the **Output measurement**.

7. ***Accept*** the default **Z factor** of **1** (*Z factor will be discussed further in Lesson 4*)

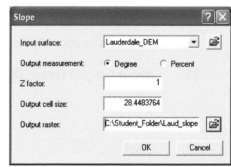

8.   *Enter* an **Output cell size** of **28.44837635**.
     (Remember that this is the grid size of the DEM.)

9.   *Save* the **Output raster** as   **Laud_slope** in your **student folder**.

     The act of saving the raster to your student folder will make the created raster permanent.

10. *Click* ____OK____. The slope grid is added to the map.

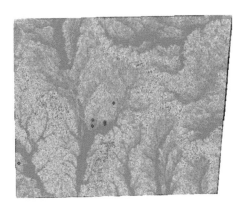

Look at the lower left hand corner of the screen to see the **processing time indicator bar.**   It might take a few minutes.

Since it is the relative value of slope of each cell that is important when calculating the cost weighted distance to each hospital we will reclassify this slope grid using a scale of 1-10. This means that the **actual degree** of slope is no longer relevant, but the **relative slope** of one cell to another is relevant. In this scenario, 1 means that the cell has the least slope; 10 meaning it has the greatest.

Another advantage of reclassifying the slope data is to reduce processing time by converting each cell slope value to an integer from a real number that can have up to five decimal places. When the absolute value of the slope is not as relevant as its relative value, this reclassification is useful.

11. To perform this reclassification, *select* **Reclassify...** from the Spatial Analyst ▼ drop-down menu.

12. *Select* **Laud_slope** as the **input raster** for the reclassification.

13. *Click* Classify... and set that the number of classes to **10**.

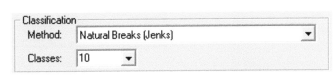

14. *Click* ____OK____. The ranked values will already be entered as **new values** in the **Reclassify** dialog box.

---

15. At the bottom of the dialog box, **save** the **output raster** as **Laud_Slope_RC** in your **student folder**.

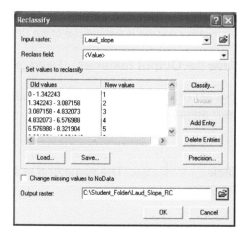

16. **Click** [OK]. The reclassified grid appears in the map display.

17. **Select Options...** from the Spatial Analyst ▼ drop-down menu.

18. **Click** the **Extent** tab.

19. **Set** the **Analysis extent** as **Same as Layer "Laud_slope"**.

20. **Click** [OK].

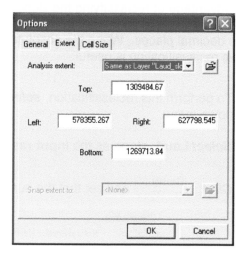

21. To perform the distance analysis, *select* **Distance ▶ Cost Weighted...** from the Spatial Analyst ▼ drop-down menu. *Set* the **Distance to** layer as **Lauderdale_Hospitals**

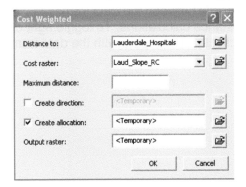

22. *Set* **Cost raster** as **Laud_Slope_ RC** (because the cost here will be associated with the "cost" of traveling the terrain represented by the reclassified slope grid).

23. *Click* the **checkbox** to **Create allocation**. **Allocation** simply identifies which cells belong to which source. In this case, it will identify which cells are associated with which hospital. You may leave **both of the output rasters as temporary**.

24. *Click* ___OK___.

25. **Look** and the lower left hand corner of the screen to see the **processing time indicator bar.** It will take a few minutes to complete this operation. When the calculation is complete, the cost weighted distance grid appears in the map display.

In this scenario, this new layer displays the distance in bands extending approximately 5,100 meters each from every hospital location in the county.

26. *Uncheck* the **CostDistance** layer in the Table of Contents to view the **CostAllocation** layer. The **CostAllocation** identifies for each cell which source (hospital) is the least costly to reach, in this case, based on the cost of slope.

27. *Turn* the **CostDistance** layer back on in the Table of Contents.

28. To compare the cost weighted distance to hospitals with the straight line distance you calculated earlier, *right click* the **Straight Line Distance** data frame and *select* **Activate**. Make certain the Hosp_Dist_SL layer is on.

29. *Toggle* between the data frames to view the differences between the two different distance calculations. To toggle right click on the frame you want to view and click activate. The repeat with the data frame.

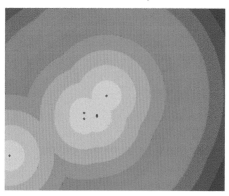

Straight Line Distance to Hospitals

Cost Weighted Distance to Hospitals

Notice that while the distance patterns are similar, the **CostWeighted** grid surface shows greater access variation based on cost constraints that were not considered in the straight line distance analysis. Distance can be calculated based on other cost factors as well to more realistically determine access to sites such as hospitals, schools, shopping centers, and tourist locations.

30. *Open* **Layer Properties** for the **CostDistance** data layer that you created.

31. *Click* the **Source** tab.

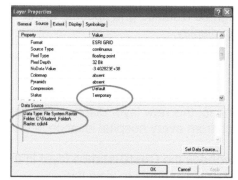

32. *Scroll* down to the *Status* property and notice that it is specified as **Temporary**. (Also, notice that this file is contained in your student folder as you specified at the beginning of this exercise.)

---

33. *Close* Layer Properties.

34. *Right click* the **CostDistance** data frame and *make this layer permanent* and then **Save** as **Hosp_Dist_CW** in your student folder.

35. *Right click* the **Cost Allocation** data frame and *make these layers permanent*. **Save** as **Hosp_Alloc** in your student folder.

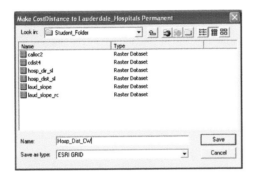 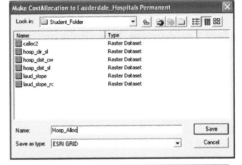

36. *Open* Layer Properties for the **CostAllocation** layer again. Notice that the status of this data file is now **permanent**. By saving the map document all temporary data files became permanent.

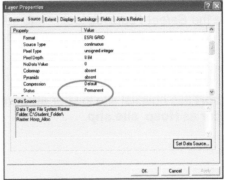

37. *Save* the project. Allow your instructor to see your work.

Cost Weighted Distance to Single Source

Additional distance allocation surfaces can be generated considering only a single source location. In this case, the **Cost Weighted Distance** is calculated from a hospital site that is located in the **Cost Allocation Zone** of a potential assisted living facility for the elderly.

1. In the Table of Contents for the **Cost Weighted Distance** data frame, make sure that the **Lauderdale_Hospitals** and **CostAllocation** are on and all other layers are turned off.

2. *Add* ✛ the **Lauderdale_Site.shp** layer the **C:\ STARS\ExtToolsinSA** folder. This potential site for a new assisted living facility for the elderly. from is a

Notice the allocation region in which the site is located.

You will now select the hospital that the site is allocated to and create a new shapefile containing only that source site.

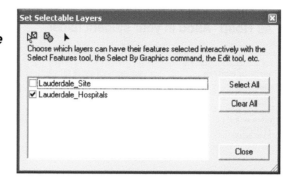

3.   From the **Selection** menu, *choose* **Set Selectable Layers**

4.   **Select Lauderdale_Hospitals** as the only selectable layer.

5.   *Click* **Close**.

6.   Use the *Select Features* tool to select the hospital to which the site is allocated.

7.   *Export* this hospital site to your **student folder** as **Hosp_site.shp**.

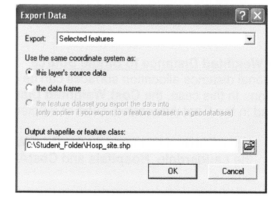

8.   When prompted, *add* the new layer to the map display.

9.   *Turn off* the **Lauderdale_Hospitals** layer in the Table of Contents.

10.   To determine the Cost Weighted Distance to this one (1) site, *select* **Distance▶Cost Weighted...** from the Spatial Analyst ▼ drop-down menu.

11.   *Select Hosp_*site as the **Distance to layer**

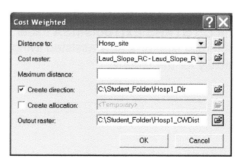

12. **Select Laud_Slope_RC** as the **Cost raster**.

13. **Choose** to **Create direction** and **save** that raster in your **student folder** as **Hosp1_Dir**.

14. **Save** the **output raster** in your **student folder** as **Hosp1_CWDist**.

15. **Click** [OK]. The grids will add to the map display. This will take a few minutes to calculate.

These maps display the **Cost Weighted Distance** and **Direction** from the one single hospital site rather than from all of the sites in the county. It is relevant when determining the cost distance to this site from a single potential assisted living facility in the same cost allocation zone. The **Direction** displayed in this scenario shows the direction, in **Right-Left** classes, for each cell to the selected hospital site given the constraints, or costs, of slope.

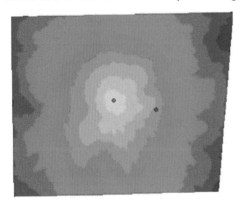 

16. **Save** the project.

Shortest Path

In addition to performing distance calculations using Spatial Analyst, you can also identify the shortest path from one location to another location considering costs associated with the potential paths. This type of analysis is particularly helpful for applications such as finding the best location and route for a new road. For our analysis, we will identify the best (shortest) route from our site of an assisted living facility to the closest hospital. This path may become a new emergency express route.

To find the shortest path from the site to the hospital,

1. **Select Distance/Shortest Path...** from the Spatial Analyst ▼ drop-down menu.

2. **Set Lauderdale_Site** as the **Path to** layer.

3. **Select Laud_Slope_RC** as the **Cost distance raster**

4. **Set Hosp1_Dir** as the **Cost direction raster**.

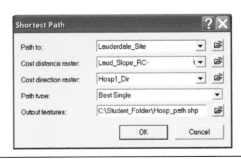

5.   *Select* **Best Single** as the **Path type**. **Best Single** path type will determine only one path (the least cost). Other path type options include:

Path Type	Output
For Each Cell	Determines a path for each cell in each zone
For Each Zone	Determines the least cost path for each zone

6.   *Save* the **Output feature** as **Hosp_path.shp** in your **student folder**.

7.   *Click* [OK]. The least cost path will appear in the map display.

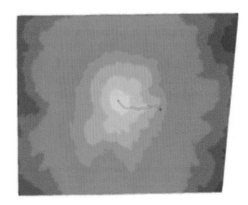

8.   *Click* the symbol for the **Hosp_path** layer in the Table of Contents to open the **Symbol Selector**.

9.   *Change* the **width** of the line to **2**.

The least cost path to the hospital will now be easier to see in the map display.

<u>Creating the Map Layout:</u>
You will now create a map layout to display the analysis results from the lesson activities.

1.   *Switch* to **Layout View** by clicking on the **Layout View** ⊙ ◻ button at the bottom of the map display.

2.   To change the page orientation of the layout page, *select* **Page and Print Setup...** from the **File** menu and *set* the **Page Orientation** to **Landscape**.

3.   *Click* [OK].

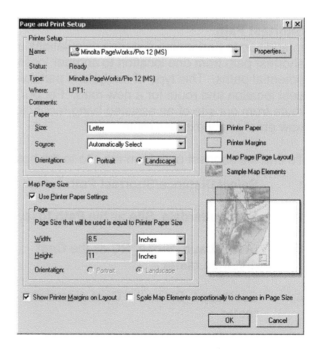

The data frames will be stacked on top of one another on the layout page.

4.   *Position* the data frame boxes by dragging them to new locations on the layout using your mouse.

5.   *Resize* the data frames on the map layout page by clicking on the data frame box and using the sizing handles to click and drag.

6.   To place a title on the layout page, *select* **Title** from the **Insert** menu. When the default title is placed on the layout page, enter the title **Mapping Distances**.

7.   *Press* **Enter** on your keyboard to accept the title.

8.   To change the font size for the title, *double click* the **title** to open **Properties**.

9.   *Click* the [Change Symbol...] button to open the **Symbol Selector**.

10.   *Change* the font **size** to **24** with **bold** **B** style.

11.   *Click* [OK] to close the **Symbol Selector**.

12.   *Click* [OK] in **Properties** to apply the change to the title.

13.   *Click* on the **Straight Line Distance** data frame.

14.   *Select* **Legend...** from the **Insert** menu. The **Legend Wizard** dialog box will open.

15.   Make sure that the **Lauderdale_Hospitals** and **Hosp_Dist_SL** layers are included as **Legend Items**.

16.   In **Legend Properties** click on 'Style...', and select the legend item on the upper left.

17.   When the legend is added to the map layout, *resize* it and *move* it to an appropriate place on the layout page near the **Straight Line Distance** data frame.

18.   *Click* on the **Cost Weighted Distance** data frame.

19.   *Select* **Legend...** from the **Insert** menu. The **Legend Wizard** dialog box will open.

20.   Make sure that the **Lauderdale_Site**, **Hosp_site**, **Hosp_path**, and **Hosp1_CWDist** layers are included as **Legend Items**.

21.   When the legend is added to the map layout, *resize* it and *move* it to an appropriate place on the layout page near the **Cost Weighted Distance** data frame.

To make the two (2) data frames have the same extent, you must make the scales the same for both data frames.

22.   *Click* the **Straight Line Distance** data frame.

23.   *Click* in the **Map Scale** box to highlight the value in the box.

24.   *Press* the **Ctrl** and **C** keys on the keyboard to copy the value.

25.   *Click* the **Cost Weighted Distance** data frame.

        The value in the **Map Scale** box probably differs from the one for the first data frame.

26.   *Click* in the **Map Scale** box so that the value is highlighted.

27.   *Press* the **Ctrl** and **V** keys on the keyboard to paste the scale from the first data frame into the **Map Scale** box for this data frame.

28.   *Press* the **Enter** key on the keyboard to apply the new scale.

        You will now insert a scale bar. However, before you add any scale bar to the layout page, you must change the display units for the data frame to select a more appropriate unit of measure for a scale bar.

29.   *Double click* the **Straight Line Distance** data frame in the Table of Contents to open **Data Frame Properties**.

30.   *Click* the **General** tab.

31.   *Change* the **Display Units** to **Miles**.

32.   *Click* [OK] to apply the change and close **Data Frame Properties**.

Because you set the same map scale for both data frames, you will only add a single scale bar to the layout page.

33.   *Select* **Scale Bar...** from the **Insert** menu.

34. *Choose* a **scale bar style** in the **Scale Bar Selector** dialog box.

35. *Click* [OK] to add the scale bar to the layout page.

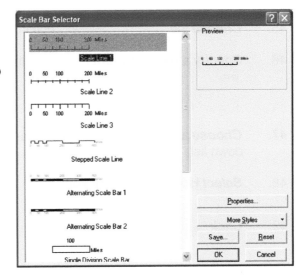

36. *Move* the **scale bar** to an appropriate place on the layout page.

You will now insert a north arrow on the layout page.

37. *Select* **North Arrow...** from the **Insert** menu.

38. *Select* a **north arrow style** in the **North Arrow Selector** dialog box.

39. *Click* [OK] to add the north arrow to the layout page.

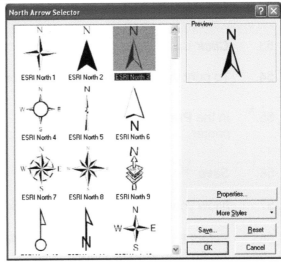

40. *Click* and *drag* the **north arrow** to an appropriate place on the layout page.

41. To place your **name** and the **date** on the layout page, *select* **Text** from the **Insert** menu. A text box will appear on the layout page (just as the title box did). It may be difficult to see the text if it is placed on another map element on the page.

42. *Type* your **name** and **date** and *press* **Enter** on your keyboard.

43. *Click* and *drag* the text box to a desired location on the map layout page.

44. To add a border to the layout page, *select* **Neatline...** from the **Insert** menu. The **Neatline** dialog box will appear.

45. **Choose** the  option.

46. **Specify** a **gap** of **1 point** for the neatline.

47. **Choose** a **border** style from the drop-down list.

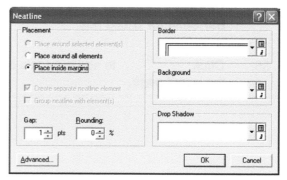

48. **Select** Hollow as the **background**.

49. **Click**  .

50. If necessary, **move** or **resize** any map elements on the layout page.

51. To export the map as an image file, **select Export Map...**from the **File** menu.

52. **Export** the layout to your **student folder** as **S3SAL1_XX** (where **XX** is your initials) in **JPEG** format.

53. **Click** Save .

54. To print the layout, **select Print...** from the **File** menu.

55. In the **Print** dialog box, **click** OK to print the map layout page to your default printer.

56. **Save** 💾 this project.

57. Allow your instructor to see your work before you exit ArcMap.

58. **Select Exit** from the **File** menu or **click** the **Close** ☒ button in the upper right corner of the ArcMap window to exit.

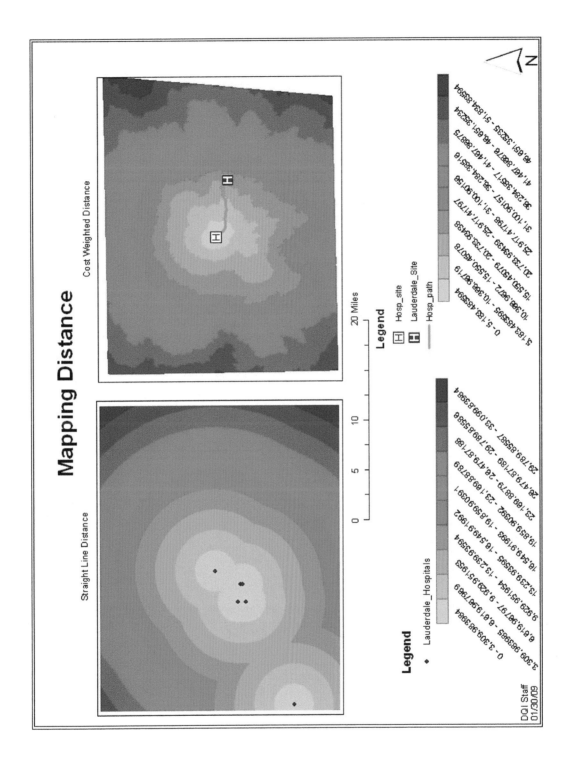

Lesson 1: Enrichment Activity

In the regular lesson activities you explored the different methods of calculating distances using Spatial Analyst. In this Enrichment Activity you will use the Cost Weighted method to determine distances to schools in your quadrangle. Since you will be using data from your own community, your maps will look different from those contained in these instructional materials, but the conceptual results will be similar.

In the regular lesson you used the slope calculation as the cost raster associated with the Cost Weighted Distance calculation. In this activity, you will use two (2) data sets to create the cost raster: land use and slope. The cost weighted analysis will determine the relative ease of access to each school location for all places in the area. In this case, if a place is a steep slope wetland area, it is relatively difficult to reach a school location from there. Conversely, if a place is below a 5% grade and is located along a transportation route, it is easy to reach a school location from there.

You may be able to think of other data sets that could be used to create the cost raster such as distance to major roads and other factors. However, for the purpose of this exercise, we will limit these factors to only land use and slope.

Cost Weighted Distance to Schools

1. *Launch* ArcMap with a **new blank map**.

2. *Save* the project as **S3SAL1_Enrich_XX.mxd** (where **XX** is your initials) in your **student folder**.

Creating the Cost Raster:

3. *Add* ✛ the **YYYY_XXX_grd** raster DEM data layer (where **XXX** is your school abbreviation and **YYYY** is your county abbreviation) from the **C:\STARS\ExtToolsinSA\local** folder.

4. *Add* ✛ the **YYYY_lulc.shp** data layer (where **YYYY** is your county abbreviation) from the **C:\STARS\ExtToolsinSA\local** folder.

Because the land use data layer is a shapefile, or feature data layer, you must convert it to a grid so that it can be used in this analysis. However, the raster DEM grid file we will use to calculate slope is projected in *UTM*, and the LULC shapefile is in *geographic* coordinates. Both files need to be set to the same spatial reference in order to perform any spatial analysis on them. Since the *geographic* coordinate system is not projected, it is not possible to measure space or distance in degrees, minutes and seconds. We will convert the LULC shapefile to the UTM projected to match the spatial reference of the raster DEM grid file.

5.   *Right click* the **YYYY_lulc.shp** data layer and *select* **Data ▶ Export Data...**

6.   *Select* the **Use the same coordinate system as the data frame** option. The data frame's coordinate system is the same as the raster DEM grid file because that file was the first one added to the new data frame.

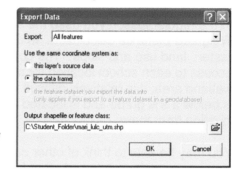

7.   *Save* the **Output shapefile** as **YYYY_lulc_utm.shp** (where **YYYY** is your county abbreviation) to your **student folder**.

8.   *Click* ⌐ OK ⌐. Note that processing may take some time due to the large number of features contained in the LULC data layer.

9.   When prompted, *click* ⌐ Yes ⌐ to add this new LULC layer to the data frame.

10.   *Right click* the **original LULC** data layer and *select* **Remove** to remove this layer from the data frame.

11.   *Save* 🖫 the project.

Now that both layers are set to the same spatial reference, we can perform spatial analysis on them to determine distances to schools. You will only need to convert the portion of the county land use data that is in your quadrangle (the extent of the DEM raster grid).

12. To set the analysis extent, **select Options...** from the Spatial Analyst ▼ drop-down menu.

13. **Click** the **Extent** tab.

14. **Set** the **Analysis extent** as **Same as Layer "YYYY_XXX_grd"**.

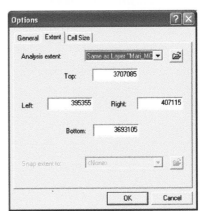

15. **Click** OK .

16. To convert the feature land use data to raster, **select Convert ▶ Features to Raster...** from the Spatial Analyst ▼ drop-down menu.

17. **Select YYYY_lulc_utm** (where **YYYY** is your county abbreviation) as the **input features layer**.

18. **Select GRIDCODE** as the **field** and

19. **Set** the **Output cell size** to **30** (since this is the cell size of each pixel in the DEM raster grid).

20. **Save** the **output raster** as **YourQuadName_lulc** in your **student folder**.

21. **Click** OK . Processing may take several minutes.

The new LULC raster grid file is added to the dataframe.

22. ***Right click*** the **YYYY_lulc_utm** data layer and ***select*** **Remove** to remove the layer from the data frame.

23. ***Zoom*** to the extent  of the remaining data layers.

24. To calculate the slope of the local quadrangle, ***select*** **Surface Analysis ▶ Slope...** from the Spatial Analyst ▼ drop-down menu.

25. ***Select*** the **YYYY_XXX_grd** data layer as the **input surface**.

26. ***Select*** **Degree** as the Output measurement to display the degree of slope generated.

27. ***Accept*** a **z factor** of **1**.

28. ***Confirm*** the **output cell size** of **30** (because this is the cell size of the input grid)

29. ***Save*** the **output raster** as **YourQuadName_Slope** in your **student folder**.

30. ***Click*** OK .

The slope raster will appear when processing is complete.

Because you are using two (2) different data sets to create a "cost" raster for this analysis, you will need to reclassify the grids to similar scales in order to specify those values that are more desirable than others in terms of costs.

First you will reclassify the land use grid that you created.

31.   To do this, *select* **Reclassify** from the Spatial Analyst ▼ drop-down menu.

32.   *Select* **YourQuadName_lulc** as the **Input raster**.

33.   *Confirm* **Value** as the **Reclass field**. The following table provides the descriptions of the various land use categories.

Gridcode	Description
11	Open Water
12	Perennial Ice/Snow
21	Low Intensity Residential
22	High Intensity Residential
23	Commercial/Industrial/Transportation
31	Bare Rock/Sand/Clay
32	Quarries/Strip Mines/Gravel Pits
33	Transitional
41	Deciduous Forest
42	Evergreen Forest
43	Mixed Forest
51	Shrubland
61	Orchards/Vineyards/Other
71	Grasslands/Herbaceous
81	Pasture/Hay
82	Row Crops
83	Small Grains

84	Fallow
85	Urban/Recreational Grasses
Gridcode	**Description**
91	Woody Wetlands
92	Emergent Herbaceous Wetlands

To reclassify the land use values, the land uses are ranked according to the cost associated with traversing them. For example, the most difficult terrain to traverse is water and wetlands. Conversely, the easiest terrain to traverse is the residential and commercial land uses, as they contain roads and other transportation networks that facilitate movement from one area to another.

34. ***Enter*** theses values in the **New Value** field. Be sure to match up the **Gridcode** with the **New Value** as listed below. Please note that **<u>not all gridcodes</u>** may exist in your database.

Gridcode	Description	New Value
0	No Data	NoData
11	Open Water	10
21	Low Intensity Residential	1
22	High Intensity Residential	1
23	Commercial/Industrial/Transportation	1
31	Bare Rock/Sand/Clay	5
32	Quarries/Strip Mines/Gravel Pits	7
33	Transitional	5
41	Deciduous Forest	5
42	Evergreen Forest	5
43	Mixed Forest	5
51	Shrubland	3
61	Orchards/Vineyards/Other	7
71	Grasslands/Herbaceous	2
81	Pasture/Hay	2
82	Row Crops	3
83	Small Grains	3
84	Fallow	3
85	Urban/Recreational Grasses	3
91	Woody Wetlands	10
92	Emergent Herbaceous Wetlands	10
If any "NoData" values appear **below** your last entry, leave as "NoData"		

35. ***Save*** the **Output raster** as **lulc_reclass** in your **student folder**.

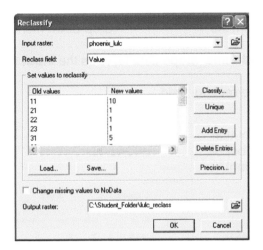

36. **Click** OK .

The reclassified land use grid will be added to the map display.

Your land use map now displays the land uses according to the *degree of difficulty to traverse code* assigned to each category.

You will now reclassify the slope grid. Keep in mind that areas with lower slopes have less cost associated with traversing them.

37. **Select** **Reclassify** from the Spatial Analyst ▼ drop-down menu.

38. **Select** the **YourQuadName_slope** data layer as the **input raster**

39. **Confirm** **Value** as the **reclass field**.

40. **Click** the Classify... button and **confirm** the number of classes is set as **10** (ten).

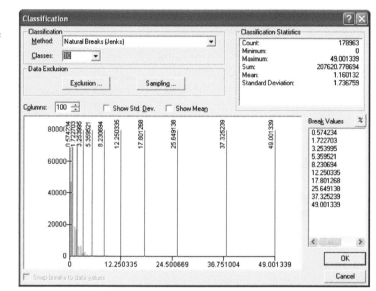

41. **Click** OK .

The reclassified values should automatically be listed from 1 to 10 from smallest to largest slope categories.

42. **Save** the **output raster** as **slope_reclass** in your **student folder**.

43. **Click** OK .

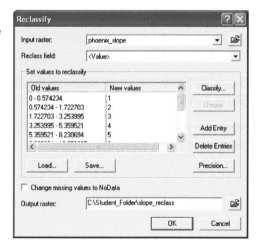

The reclassified slope data layer will be added to the map display.

Your map now displays the reclassified slope values of your quad ranging from 1-10, with 1 being the least slope, and 10 being the greatest slope.

You will now "combine" the reclassified grids to create a single grid using the **Raster Calculator**. However, we will "weight" the input grids based on the influence that they will have in determining our overall cost grid. For our purposes, we will assign a weight of 66% (0.66) to the slope grid and a weight of 34% (0.34) to the land use grid.

44.   **Select** **Raster Calculator...** from the
      Spatial Analyst ▼ drop-down menu.

45.   Using the data layer names in the **Layers list** and the **number** and **operator buttons** on the **Raster Calculator** dialog box, **enter** the following equation into the **Raster Calculator** equation box:

      [Slope_Reclass] * .66 + [LULC_Reclass] * .34

46.   **Click** Evaluate to create the cost grid.

The "combined" raster data layer resulting from the raster calculation will be added to the map display.

47.   **Open** **Layer Properties** for the new **Calculation** layer and **click** the **Symbology** tab.

48.   **Change** the **Color ramp** to ▭▭▭▭ **(Red to Green)**.

49.   **Check** to **Invert** the color ramp to show the higher values in red.

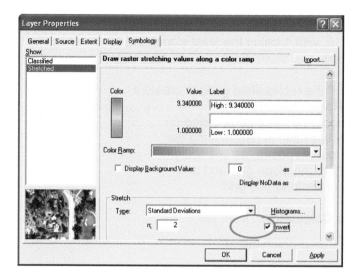

50. **_Click_** Apply .

51. **_Click_** the **Source** tab.

52. **_Scroll_** down to view the **Raster Information**. Note that the status of this layer is **temporary**.

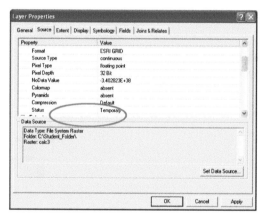

53. **_Close_** **Layer Properties**.

54. To make this layer permanent, **_right click_** the **Calculation** layer in the Table of Contents and **_select_** **Make Permanent...**

55. **_Save_** this layer as **YourQuadName_cost** in your **student folder**.

This new map displays the combined weighted cost, or difficulty to traverse, for each cell in the raster grid with respect to slope and land use. The land use and slope values for each cell are summed to arrive at a combined value for each cell. The areas in green have the lowest combined value, and therefore the lowest costs associated with them; conversely, the areas in red have the highest combined value, and hence, the highest cost, and are the most difficult to traverse.

56. **Save** 🖫 the project.

Adding the Source Layer:

1. **Add** ⬇ the **YYYY_school.shp** layer (where **YYYY** is your county abbreviation) from the **C:\STARS\ExtToolsinSA\local** folder. All of the county schools are added to the map.

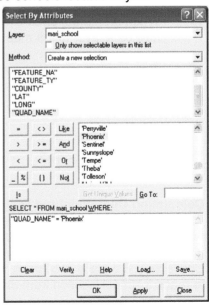

2. To select only those schools located in your local quadrangle, **choose Select by Attributes...** from the **Selection** menu.

3. Using the field names in the field list and the **operator buttons** in the **Select by Attributes** dialog box, **build** an equation to select those schools located in your quadrangle.

   "QUAD_NAME" = YourQuadName

4. **Click** Apply .

5. **Close** the **Select By Attributes** dialog box.

The schools located in your local quadrangle should be selected in the map.

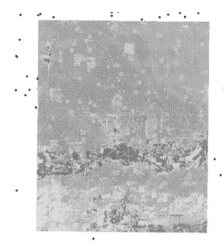

6. To save these schools as a separate data layer, *right click* the **school** data layer in the Table of Contents and *select* **Data ▶ Export Data...**

7. At the top of the **Export Data** dialog box, *confirm* that the **export** option is set to **Selected features** so that only the selected features are exported to the new shapefile.

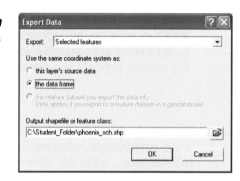

8. *Select* the option **the data frame** to use its coordinate system.

9. *Save* **Output shapefile** as **YourQuadName_sch** in your **student folder**.

10. *Click* [OK] to export the data.

11. When prompted, *click* [Yes] to add the new data layer to the map.

12. *Remove* the original county school data layer.

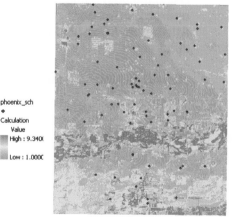

13. *Save* 🖫 the project.

Calculating Cost Weighted Distance:

This cost weighted analysis will determine the relative ease of access to each school location for all places in the area. Based on the reclassified zones set up earlier in this exercise, if a place is a steep slope wetland area, it is relatively difficult to reach a school location from there. Conversely, if a place is below a 5% grade and is located along a transportation route, it is easy to reach a school location from there.

Because the extent of the cost grid and the schools data layer probably aren't exactly the same, you will need to set the analysis extent to make sure that all cells in the cost grid are included in the distance raster. To do this,

1. **Select Options...** from the Spatial Analyst ▼ drop-down menu.

2. **Click** the **Extent** tab.

3. **Set** the **analysis extent** as **Same as Layer "Calculation"**.

4. **Click** OK .

5. **Select Distance ▶ Cost Weighted...** from the Spatial Analyst ▼ drop-down menu.

6. **Select** the **YourQuadName_sch** data layer as the **Distance to** layer.

7. **Select** the **Calculation** data layer as the **Cost raster**. (It is still named Calculation in the Table of Contents even though you made it permanent and saved it in your student folder earlier.)

8. **Save** the **output raster** as **YourQuadName_cwd** in your **student folder**.

9. **Click** OK . The distance grid is added to the map.

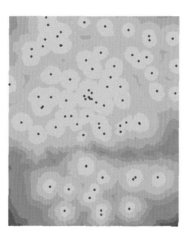

☑ phoenix_sch
  ◦
☑ phoenix_cwd
  ☐ 0 - 846.1303711
  ☐ 846.1303712 - 1,692.260742
  ☐ 1,692.260743 - 2,538.391113
  ☐ 2,538.391114 - 3,384.521484
  ☐ 3,384.521485 - 4,230.651855
  ☐ 4,230.651856 - 5,076.782227
  ☐ 5,076.782228 - 5,922.912598
  ☐ 5,922.912599 - 6,769.042969
  ☐ 6,769.04297 - 7,615.17334
  ☐ 7,615.173341 - 8,461.303711

Notice that distance from each school site is not the only relevant factor when accounting for accessibility. Slope and land use have been accounted for and determined to play a significant factor is determining site accessibility. This new layer displays the distance in bands extending 1,600 meters each from every school site in the area. It is more difficult to travel to those school sites from those areas within the darker bands with respect to slope and land use than those located within the lighter bands.

10. **Uncheck** all of the layers in the Table of Contents except the **YourQuadName_cwd** and **YourQuadName_sch** data layers.

11. **Save** 💾 the project.

Creating the Map Layout:
You will now create a map layout to display the analysis results from this enrichment activity.

1. **Switch** to **Layout View** by clicking on the **Layout View** button at the bottom of the map display.

2. To change the page orientation of the layout page, **select Page and Print Setup...** from the **File** menu. **Set Page Orientation** to **Landscape**.

3. **Click** OK.

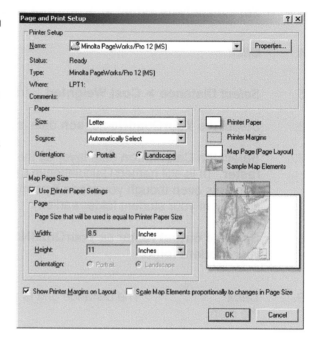

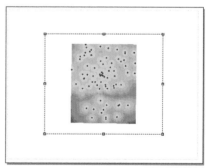

3. **Resize** the data frame on the map layout page by clicking on the data frame box and using the sizing handles to click and drag.

4. **Position** the data frame box by dragging it to a new location on the layout using your mouse.

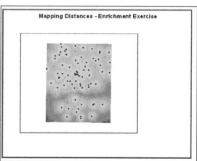

Mapping Distances - Enrichment Exercise

5. To place a title on the layout page, **select Title** from the **Insert** menu. When the default title is placed on

the layout page, enter the title **Mapping Distances – Enrichment Exercise**.

6.   *Press* **Enter** on your keyboard to accept the title.

7.   To change the font size, *double click* the **title** to open **Properties**.

8.   *Click* the 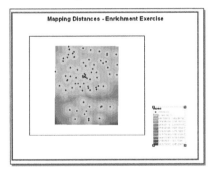 button to open the **Symbol Selector**.

9.   *Change* the font **size** to **24** with **bold** **B** style.

10.   *Click* **OK** to close the **Symbol Selector**.

11.   *Click* **OK** in **Properties** to apply the change to the title.

14.   *Select* **Legend...** from the **Insert** menu. The **Legend Wizard** dialog box will open.

15.   Make sure that the **YourQuadName_sch** and **YourQuadName_cwd** layers are included as **Legend Items**.

16.   When the legend is added, *move* it to an appropriate place on the layout page.

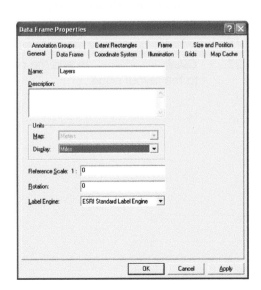

You will now insert a scale bar. However, before you add any scale bar to the layout page, you must change the display units for the data frame. The display units are currently set as meters. Miles is a more appropriate unit of measure for a scale bar.

17.   *Double click* the **Layers** data frame in the Table of Contents to open **Data Frame Properties**.

18.   *Click* the **General** tab.

19.   *Change* the **Display Units** to **Miles**.

20.   *Click* **OK** to apply the change and close **Data Frame Properties**.

21.   *Select* **Scale Bar...** from the **Insert** menu.

22.   *Choose* a **scale bar style** in the **Scale Bar Selector** dialog box.

23.   *Click* **OK** to add the scale bar to the layout page.

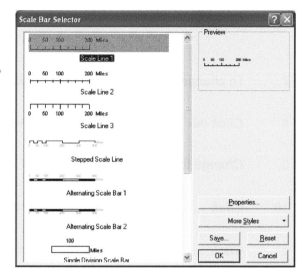

24.   *Move* the **scale bar** to an appropriate place on the layout page.

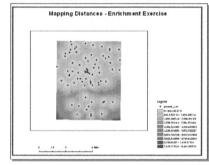

You will now insert a north arrow on the layout page.

25.   *Select* **North Arrow...** from the **Insert** menu.

26.   *Select* a **north arrow style** in the **North Arrow Selector** dialog box.

27.   *Click* **OK** to add the north arrow to the layout page.

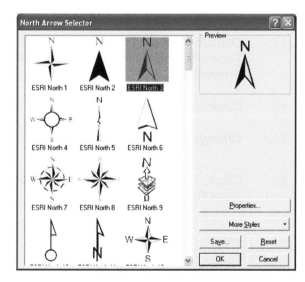

28. ***Click*** and ***drag*** the **north arrow** to an appropriate place on the layout page.

29. To place your **name** and the **date** on the layout page, ***select*** **Text** from the **Insert** menu. A text box will appear on the layout page (just as the title box did). It may be difficult to see the text if it is placed on another map element on the page.

30. ***Type*** your **name** and **date** and ***press*** **Enter** on your keyboard.

31. ***Click*** and ***drag*** the text box to a desired location on the map layout page.

32. To add a border to the layout page, ***select*** **Neatline...** from the **Insert** menu. The **Neatline** dialog box will appear.

33. ***Choose*** the [Place inside margins] option.

34. ***Specify*** a **gap** of **1 point** for the neatline.

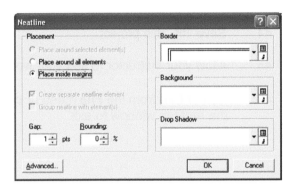

35. ***Choose*** a **border** style from the drop down list.

36. ***Select*** **Hollow** as the **background**.

37. ***Click*** [OK].

38. If necessary, ***move*** or ***resize*** any map elements on the layout page.

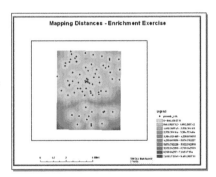

---

39.   To export the map as an image file, *select Export Map...*from the *File* menu.

40.   *Export* the layout to your **student folder** as **S3SAL1_Enrich_XX** (where **XX** is your initials) in **JPEG** format.

41.   *Click* [Save].

42.   To print the layout, *select Print...* from the *File* menu.

43.   In the **Print** dialog box, *click* [OK] to print the map layout page to your default printer.

44.   *Save* 💾 this project.

45.   Allow your instructor to see your work before you exit ArcMap.

46.   *Select* **Exit** from the *File* menu or *click* the **Close** [X] button in the upper right corner of the ArcMap window to exit.

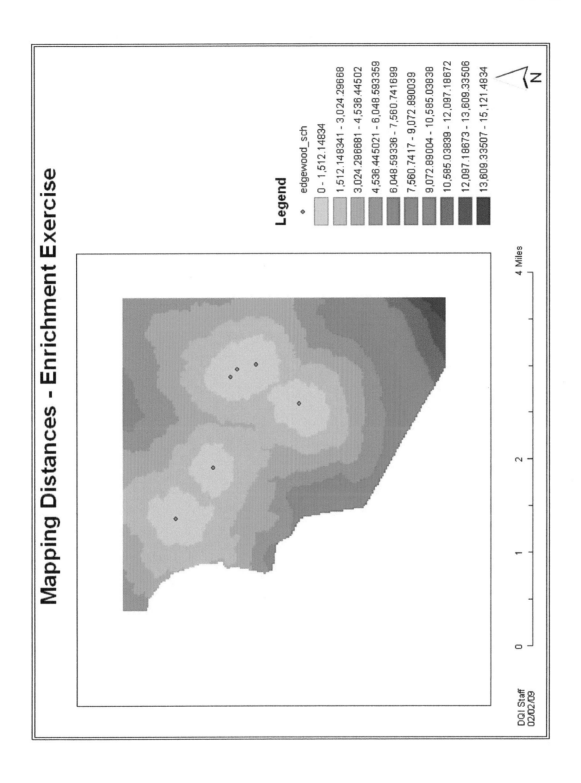

Mapping Distances - Enrichment Exercise

Legend

◇ edgewood_sch

0 - 1,512.14834

1,512.148341 - 3,024.29668

3,024.296681 - 4,536.44502

4,536.445021 - 6,048.593359

6,048.59336 - 7,560.741699

7,560.7417 - 9,072.890039

9,072.89004 - 10,585.03838

10,585.03839 - 12,097.18672

12,097.18673 - 13,609.33506

13,609.33507 - 15,121.4834

0   1   2   4 Miles

DQI Staff
02/02/09

Lesson 1: Mapping Distance Lesson Review

Key Terms
Use the lesson or index provided in the back of the book to define each of the following terms.

1. Straight line distance
2. Cost-weighted distance
3. Euclidean Distance
4. Direction grid
5. Cost Allocation

Global Concepts
Use the information from the lesson to answer the following questions. You may need to answer these on the back of this page or on your own paper.

6. The distances calculated in previous exercises are different from your calculated distances from this exercises because what factor was taken into account?

7. There are many types of 'costs' that can be considered in cost-weighted distance – please list three and give examples.

8. Why would you use straight line distance measurements instead of cost-weighted?

Let's Talk About It...
Answer the following question and share the responses with your instructor and classmates.

9. With the addition of these skills to your skill set, what would you be better able to describe or analyze now? Give an example other than what is in the exercise.

Lesson 1

Lesson 2:
Mapping Density

Another way that spatial analysis is useful is to measure the density of features. The density function provides the measure of a quantity of certain point or line features over a given unit of area (for example, population per square mile). A density map is often more useful than representing data over a wide area with a single point or multiple points. In the example below, the figure on the left shows select United States cities represented with graduated points corresponding to populations. The figure on the right shows the same select United States cities using density to represent population per square mile.

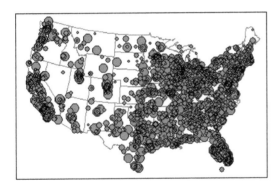 

ArcGIS features three density types: point density, kernel density, and line density.

Point data about people and places displays information about these features at a specific location in the form of attributes. At certain scales, this information is useful to see the distribution or pattern over an area; however as with example above using the point shapefile can result in a map that is difficult to read. In order to better represent the populations, a population density raster can be created that shows how the population is distributed around population points. In this lesson, you will explore the concept of density calculation in Spatial Analyst. We will calculate density using Simple and Kernel calculations.

In a simple or point density calculation, points that fall within the search area are summed and then divided by the search area size to get each cell's density value. The kernel density

calculation weights cells within a search radius according to their proximity to each point, which differs from point density in that each point was considered equal.   Kernel density assumes that points further away are relevant but less influential than closer points. Points near the center of a search area are weighted more heavily than those near the edge. Cells at the center in the kernel grid surface will have a higher value than those at the center of the simple grid surface. Differences in scale and application determine when one process is preferred over another. Generally speaking, the simple density is used for larger scale maps, and when the data is lower quality. The kernel density better pinpoints the density, so is used with better data quality and in smaller scale maps.

Lesson 2: Mapping Density

Sometimes it is more helpful to see how point values are distributed over a surface instead of encapsulated as individual points. For instance, populations of cities are not all in one point – they are distributed over an area. In order to better represent the populations, a population density can be created that shows how the population is distributed around population points. In this lesson, you will explore the concept of density calculation in Spatial Analyst.

Calculating Density

1.   *Launch* **ArcMap** with a **new blank map** and *save* it as **S3SAL2_XX** in your student folder (where **XX** is your initials).

2.   **Add** the **US_Cities_Cont** and **US_States_Cont** data layers located in the **C:\STARS\ExtToolsinSA** folder.

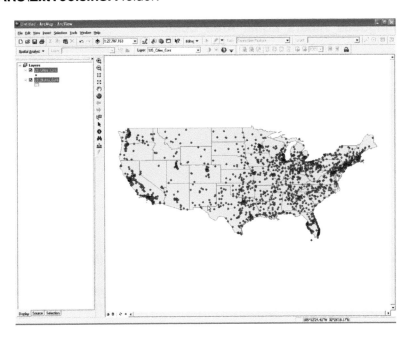

Because the map units for this data are currently in degrees minutes seconds (DMS), they must be changed to a measurable projected coordinate system because the density calculation is computed in square area.

3.   ***Right click*** **Layers**.

4.   ***Open*** **Properties**.

5.   ***Click*** the **Coordinate System** tab.

6.   ***Select*** **Predefined**.

7.   ***Select*** **Projected Coordinate Systems**.

8.   ***Select*** **Continental**.

9.   ***Select*** **North America**.

10.   ***Select*** **USA_Contiguous_Albers_Equal_Area_Conic**.

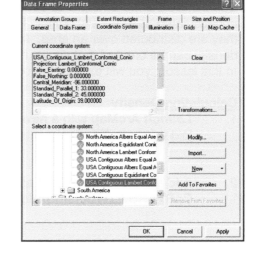

11.   ***Click*** `OK` .

 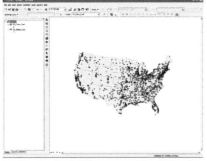

Not Projected (Decimal Degrees)   **Projected (Measured in Miles)**

The map now changes to display into a measurable coordinate system, but the data layers themselves are not yet projected to where density calculations can be made.

12.   ***Right click*** **US_Cities_Cont**.

13.   ***Select*** **Data**.

14.   ***Select*** **Export Data**.

15.   ***Use the same coordinates system as:*** the data frame .

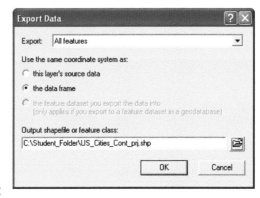

16.   ***Save*** the **Output shapefile or feature class** as **US_Cities_Cont_prj.shp** in your **student folder**.

---

17. **Click** 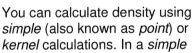.

18. **Add** this exported data to the map as a layer.

19. **Repeat** this process for **US_States_Cont.shp.**

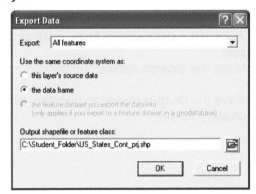

20. **Remove** the original **Cities (Us_Cities_Cont)** and **States (US_States_Cont)** layers.

21. **Rename** the data frame **Kernel Density**.

The map now displays the new layers in a projected coordinate system that is measurable so that density calculations can be made.

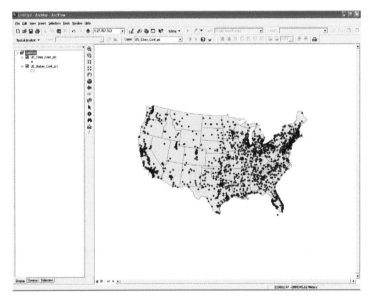

22. **Select Density...** from the Spatial Analyst ▼ drop-down menu.

23. **Confirm** that **US_Cities_Cont_prj** is set as the **Input data**.

24. **Confirm Population field** is set as **POP1990.**

You can calculate density using *simple* (also known as *point*) or *kernel* calculations. In a *simple* or *point* density calculation, points that fall within the search area are summed and then divided by the search area size to arrive at each cell's density value. This gives a simple and quick estimation about the density, and takes up less processing time. It is less of an issue now with the speed of computers, but earlier this process could take hours, and so this process was necessary to determine if it was worth the effort to do a kernel density.

The *kernel* density calculation works similarly to the *simple* or *point* density calculation, except it weights cells within a search radius according to their proximity to each point; cells lying near the center of a search area are weighted more heavily than those lying near the edge. The result is a tighter smoother distribution of values.

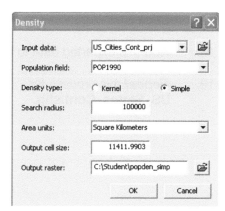

25. For this first density calculation, *select* simple as the density type.

26. *Accept* the **defaults** for **Area units** and **Output cell size**.

27. Make the *Search radius* 100,000

28. *Save* the **Output raster** as **popden_simp** in your **student folder**.

29. *Click* [OK] .

30. *Turn off* the **Cities** data layer in the Table of Contents so that you can see the population density grid layer. Change the **States** data layer to **hollow**, **30% gray** with a **width** of 0.4 to give context.

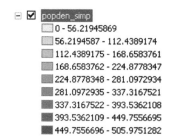

This new map displays a density surface that shows the number of people per square kilometer in the area around each grid cell across the continental United States. The lightest color represents areas with lower density population, from 0-56 people per square kilometer, ranking up to the darkest areas where there are 449 to 506 people per square kilometer.

31. For this second density calculation, *select* Kernel as the density type.

32. *Accept* the **defaults** for **Area units** and **Output cell size**.

33. Change the default for **Search radius** to 100,000

34. *Save* the **Output raster** as **popden_kern** in your **student folder**.

35. *Click* [OK] .

---

36. **Turn on** the **popden_kernel** data layers in the Table of Contents so that you can see the population density grid layer.

This new map displays a density surface that shows the number of people per square kilometer in the area around each grid cell across the continental United States. The lightest color represents areas with lower density population, from 0-127 people per square kilometer, ranking up to the darkest areas where there are 1,018-1,145 people per square kilometer.

For your third exercise, you will look at kernel density with half the area for each search radius

37. **Select** **Kernel** density type.

38. **Accept** the **defaults** for **Area units** and **Output cell size**. Set the **Search Radius** to 50,000

39. **Save** the **Output raster** as **PopDen_ker5** in your **student folder**.

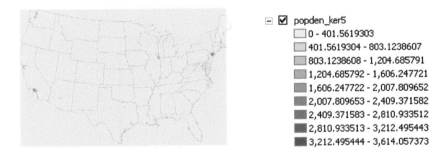

This new map displays a density surface that also shows the number of people per square kilometer across the continental United States. The same type of range results, where the lightest color represents areas with lower density population, in this case from 0-401 people per square kilometer, ranking up to the darkest areas where higher population densities exist, in this case 3,212 to 3.614 people per square kilometer.

40. **Zoom in** to a specific area and **toggle** between the three density surfaces to see the differences between them.

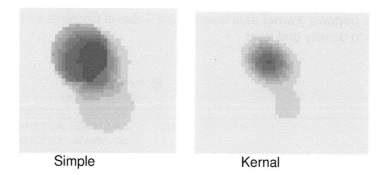

Simple   Kernal

41. **Use** the **Identify Tool** 🛈 to reveal the grid values of the same cells in the **simple** and **kernel** density surfaces.

42. **Click** on cells toward the center of the density concentration.

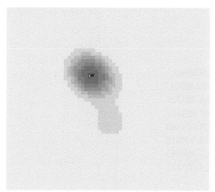

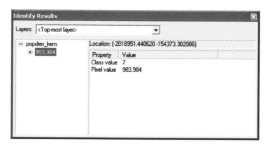

43. **Click** cells outward toward the perimeter.

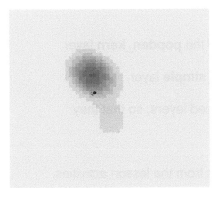
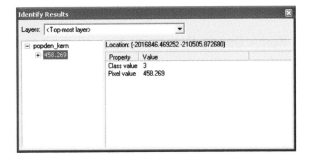

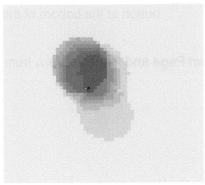
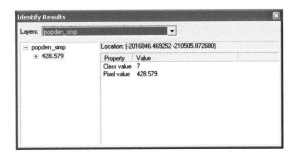

Notice that the cells at the center in the **kernel** grid surface have a much higher population density value than those at the center of the **simple** grid surface. Also notice that the cells out toward the perimeter of the density concentration are also higher for the **kernel** surface than the **simple** surface, but the difference is not as great. This is because the **kernel** calculation weights the cells according to their proximity to the points in their search radius. So, in this case, the cells at the center of the distribution will be weighted more heavily than those away from the center. The resulting grid surface will be a tighter smoother surface when using the kernel calculation. Differences in scale and application determine when one process is preferred over another.

44. **Save** the **project**.

45. In order to display all three maps in one layout, **switch** to **Layout view** .

46. **Right click** on the Layers data frame.

47. **Select copy**.

48. **Right click** within the layout,

49. **Select paste**.

50. **Right click** within the layout,

51. *Select* **paste** again

52. *Name* the first data frame **Simple Density** and *turn off* the **popden_kern** layer.

53. In the Kernal Density Data frame *turn off* the **popden_simple** layer.

53. In the final Data Frame, turn off both previously mentioned layers, so that they **popden_ker5** layer is visible.

Creating the Map Layout:
You will now create a map layout to display the analysis results from the lesson activities.

1. *Switch* to **Layout View** by clicking on the **Layout View** ▣ button at the bottom of the map display.

2. To change the page orientation of the layout page, *select* **Page and Print Setup...** from the **File** menu.

3. *Set* the **Page Orientation** to **Portrait**.

4. *Click* [OK].

   The data frames will be stacked on top of one another on the layout page.

5. *Position* the data frame boxes by dragging them to new locations on the layout and *resize* the data frames on the map layout page by clicking on the data frame box and using the sizing handles to click and drag.

6. To place a title on the layout page, *select* **Title** from the **Insert** menu. When the default title is placed on the layout page, enter the title **Kernel vs. Simple Density**.

7. *Press* **Enter** on your keyboard to accept the title.

8. To change the font size for the title, *double click* the **title** to open **Properties**.

9. *Click* the [Change Symbol...] button to open the **Symbol Selector**.

10. *Change* the font **size** to **24** with **bold** **B** **style**.

11. *Click* [OK] to close the **Symbol Selector**.

12. *Click* [OK] in **Properties** to apply the change to the title.

13. *Click* on the **Kernel Density(radius 100,000)** data frame.

14. *Select* **Legend...** from the **Insert** menu. The **Legend Wizard** dialog box will open.

15.   Make sure that the **popden_kern** layer is included as a **Legend Item**.

16.   When the legend is added to the map layout, *resize* it and *move* it to an appropriate place on the layout page near the **Kernel Density** data frame.

17.   *Click* on the **Simple Density(radius 100,000)** data frame.

18.   *Select* **Legend...** from the **Insert** menu. The **Legend Wizard** dialog box will open.

19.   Make sure that the **popden_simple** layer is included as a **Legend Item**.

20.   *Click* on the **Kernal Density(radius 50,000)** data frame.

21.   *Select* **Legend...** from the **Insert** menu. The **Legend Wizard** dialog box will open.

22.   Make sure that the **popden_ker5** layer is included as a **Legend Item**.

23.   When the legend is added to the map layout, *resize* it and *move* it to an appropriate place on the layout page near the **Simple Density** data frame.

        You will now insert a scale bar. However, before you add any scale bar to the layout page, you must change the display units for the data frame to select a more appropriate unit of measure for a scale bar.

24.   *Double click* the **Kernel Density** data frame in the Table of Contents to open **Data Frame Properties**.

25.   *Click* the **General** tab.

26.   *Change* the **Display Units** to **Miles**.

27.   *Click* OK to apply the change and close **Data Frame Properties**.

        Because you set the same map scale for both data frames, you will only add a single scale bar to the layout page.

28.   Repeat steps 24-27 for the other two data frames.

29.   *Select* **Scale Bar...** from the **Insert** menu.

30.   *Choose* a **scale bar style** in the **Scale Bar Selector** dialog box.

31.   *Click* OK to add the scale bar to the layout page.

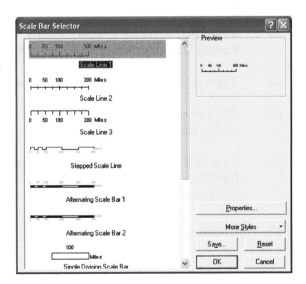

32. Repeat steps 29-31 for the other two data frames.

33. ***Move*** the **scale bars** to an appropriate places on the layout page.

You will now insert a north arrow on the layout page.

34. ***Select* North Arrow...** from the **Insert** menu.

35. ***Select*** a **north arrow style** in the **North Arrow Selector** dialog box.

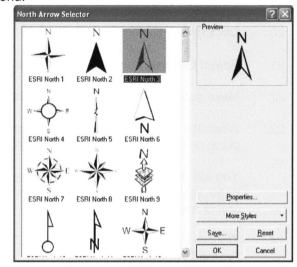

36. ***Click*** [OK] to add the north arrow to the layout page.

37. ***Click*** and ***drag*** the **north arrow** to an appropriate place on the layout page.

38. To place your **name** and the **date** on the layout page, ***select* Text** from the **Insert** menu. A text box will appear on the layout page (just as the title box did). It may be difficult to see the text if it is placed on another map element on the page.

39. ***Insert*** text to properly label the data frames as **Kernel Density(radius 100,000) , Kernel Density(radius 50,000)**and **Simple Density(radius 100,000)**.

40. ***Type*** your **name** and **date** and ***press*** **Enter** on your keyboard.

41. ***Click*** and ***drag*** the text box to a desired location on the map layout page.

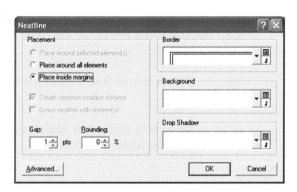

42. To add a border to the layout page, ***select*** **Neatline...** from the **Insert** menu. The **Neatline** dialog box will appear.

43. ***Choose*** the [Place inside margins] option.

44. ***Specify*** a **gap** of **1 point** for the neatline.

45.   *Choose* a **border** style from the drop-down list.

46.   *Select* **Hollow** as the **background**.

47.   *Click* [OK].

48.   If necessary, *move* or *resize* any map elements on the layout page.

49.   To export the map as an image file, *select Export Map...*from the *File* menu.

50.   *Export* the layout to your **student folder** as **S3SAL2_XX** (where **XX** is your initials) in **JPEG** format.

51.   *Click* [Save].

52.   To print the layout, *select Print...* from the *File* menu.

53.   In the **Print** dialog box, *click* [OK] to print the map layout page to your default printer.

54.   *Save* 💾 this project.

55.   Allow your instructor to see your work before you exit ArcMap.

56.   *Select* **Exit** from the **File** menu or *click* the **Close** [X] button in the upper right corner of the ArcMap window to exit.

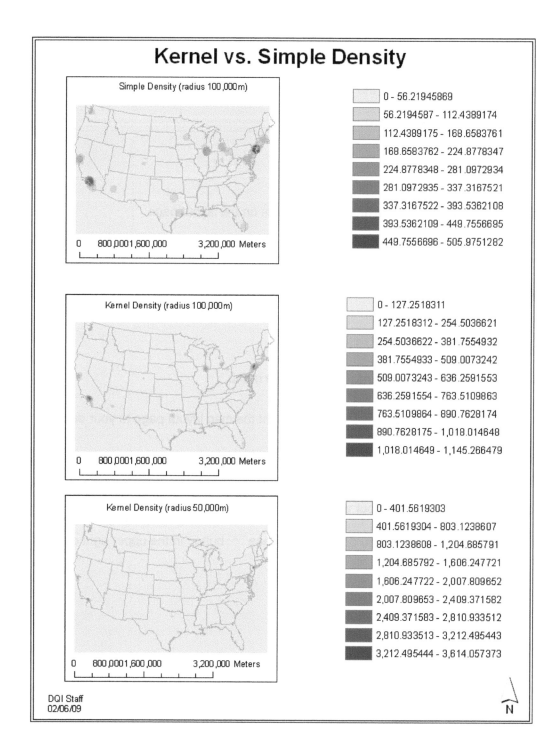

Kernel vs. Simple Density

Simple Density (radius 100,000m)

0 800,000 1,600,000 3,200,000 Meters

- 0 - 56.21945869
- 56.2194587 - 112.4389174
- 112.4389175 - 168.6583761
- 168.6583762 - 224.8778347
- 224.8778348 - 281.0972934
- 281.0972935 - 337.3167521
- 337.3167522 - 393.5362108
- 393.5362109 - 449.7556695
- 449.7556696 - 505.9751282

Kernel Density (radius 100,000m)

0 800,000 1,600,000 3,200,000 Meters

- 0 - 127.2518311
- 127.2518312 - 254.5036621
- 254.5036622 - 381.7554932
- 381.7554933 - 509.0073242
- 509.0073243 - 636.2591553
- 636.2591554 - 763.5109863
- 763.5109864 - 890.7628174
- 890.7628175 - 1,018.014648
- 1,018.014649 - 1,145.266479

Kernel Density (radius 50,000m)

0 800,000 1,600,000 3,200,000 Meters

- 0 - 401.5619303
- 401.5619304 - 803.1238607
- 803.1238608 - 1,204.685791
- 1,204.685792 - 1,606.247721
- 1,606.247722 - 2,007.809652
- 2,007.809653 - 2,409.371582
- 2,409.371583 - 2,810.933512
- 2,810.933513 - 3,212.495443
- 3,212.495444 - 3,614.057373

DQI Staff
02/06/09

N

Lesson 2: Enrichment Activity

You will now extend the concepts that you just learned pertaining to using the Density functions of the Spatial Analyst extension to your local data. You will perform two (2) population density calculations using your state data. You will then calculate density of a set of features within your county.

Calculating Population Density

1. **Launch** ArcMap with a **new blank map**. If ArcMap is still in **Layout View**, *click* to change to **Data View** .

2. **Save** the project as **S3SAL2_Enrich_XX.mxd** (where **XX** is your initials) in your **student folder**

3. **Add** the **STATES.shp** data layer located in the **C:\ STARS\ExtToolsinSA\local** folder.

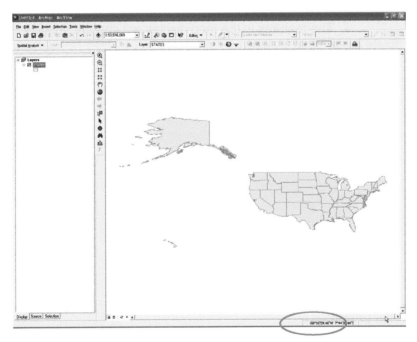

Because the map units for this data are currently in degrees minutes seconds (DMS), they must be changed to a measurable projected coordinate system because the density calculation is computed in square area. You will change the coordinate system for this data frame to the **USA Contiguous Albers Equal Area Conic** coordinate system because it uses **meters** as the unit of measure.

3. **Right Click** the **Layers** data frame in the Table of Contents and *select* **Properties**.

---

4. **Click** the **Coordinate System** tab.

5. In the **Select a coordinate system** area, **click** the ⬚ Predefined coordinate system folder to view the subfolders contained in it.

6. **Click** the ⬚ Projected Coordinate Systems folder.

7. **Click** the ⬚ Continental folder.

8. **Click** the ⬚ North America folder.

9. **Select** the **USA Contiguous Albers Equal Area Conic** coordinate system. (*Note: There is also a* USA Contiguous Albers Equal Area Conic USGS *coordinate system. Be careful to choose the correct coordinate system.*)

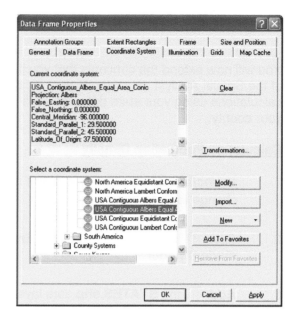

10. **Click** [OK]. The features in the map display will change based on the change to the coordinate system that is used.

11. **Save** 💾 the project.

The map now changes to display into a measurable coordinate system, but the data layers themselves are not yet projected so that density calculations can be made. You will export the data that you will need to the analysis using the coordinate system of the data frame.

12. **Select** your state on the map using the **Select Features** tool ▣.

13. **Right click** the **States** data layer and **select** Data ▶ Export Data... The **Export Data** dialog box will appear.

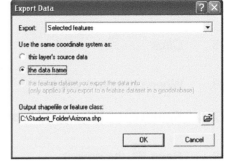

14. **Specify** to **use the same coordinate system** as ⊙ the data frame .

15. **Save** the **Output shapefile or feature class** as **YourState.shp** in your **student folder**.

16. **Click** [OK] .

17. When prompted, **click** [Yes] to add the new shapefile to the map. This new shapefile contains only the polygon of your state.

18. **Remove** the original **STATES** data layer.

19. **Zoom** to the **full extent** 🌐 of the feature in the map display.

20. **Add** ✚ the **XX_mj_cities.shp** data layer (where **XX** is your state abbreviation) located in the **C:\ STARS\ExtToolsinSA\local** folder.

Because the coordinate system of this data layer differs from the coordinate system of the data frame, the following dialog box will appear:

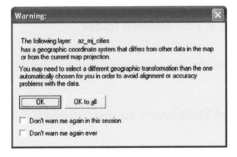

21. **Click** [OK] .

This data layer contains the location of the major cities in your state. Even though the coordinate system of this data layer differed from that of the data frame, ArcMap reprojected the layer "on-the-fly" meaning that the features in the layer could be displayed, but the original source data layer itself is not reprojected. However, the layer needs to be projected into a measurable coordinate system so that density can be calculated.

22. **Right click** the **XX_mj_cities.shp** data layer and **select Data ▶ Export Data...**

23. **Specify** to **use the same coordinates system** as the ⊙ the data frame .

24. **Save** the **Output shapefile or feature class** as **XX_mj_cities_proj.shp** in your **student folder** (where **XX** is your state abbreviation).

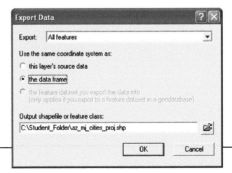

25. ***Click*** 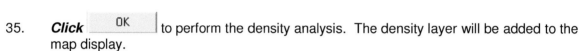 .

26. When prompted, ***click*** Yes to add the new shapefile to the map.

27. ***Remove*** the **XX_mj_cities.shp** data layer.

28. ***Save*** 💾 the project.

29. ***Select* Density...** from the Spatial Analyst ▼ drop down menu.

30. ***Select*** the **XX_mj_cities_proj** data layer as the **Input data**.

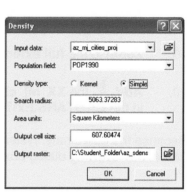

31. ***Select*** the **POP2000** field as the **Population field**.

32. ***Select*** the ⊙ Simple density option.

33. ***Accept*** the **default settings** for **Search radius**, and **Output cell size**.

34. ***Save*** the **Output raster** as **ZZ_SDens** (where **ZZ** is your state abbreviation) in your **student folder**.

35. ***Click*** 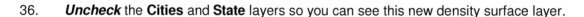 to perform the density analysis. The density layer will be added to the map display.

36. ***Uncheck*** the **Cities** and **State** layers so you can see this new density surface layer.

37. **Open Layer Properties** for the **YourState** layer.

38. **Click** the **Symbology** tab.

39. **Change** the **symbology** for the **State** layer to **no fill** with a **black**, **2-point** outline.

40. **Check** this layer in the Table of Contents so that the feature will be displayed again in the map display.

Because people tend to cluster around population centers rather than settle in a widely disbursed evenly distributed pattern throughout an area, most of the area of your state is probably in the lowest density range. To enhance the map display to show these population clusters more clearly,

41. **Open Layer Properties** for the **XX_Sdens** raster data layer.

42. If necessary, **click** the **Symbology** tab.

43. **Double click** the symbol for the **lowest category** in the layer to open a **color palette**.

44. **Click** the **No Color** option in the **color palette**.

45. **Click** ⬚ OK to apply the change and close **Layer Properties**.

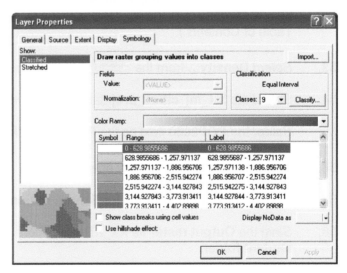

The map now displays the simple density surface grid layer for your state.

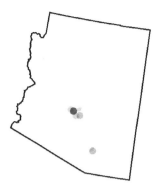

46. **Save** 💾 the project.

47. **Double click** the **Layers** data frame to open **Data Frame Properties**.

48. **Rename** the data frame **Simple Density**.

49. **Click** [OK] to close **Data Frame Properties**.

50. To insert a new data frame, **select Data Frame** from the **Insert** menu.

51. **Rename** the new data frame **Kernal Density**.

52. **Copy** your **state** and **major cities** layers from the **Simple Density** data frame into the **Kernal Density** data frame. (The **major cities** data layer will still be unchecked in the Table of Contents.)

53. **Select Density…** from the Spatial Analyst ▼ drop down menu.

54. **Select** the **XX_mj_cities_proj** data layer as the **Input data**.

55. **Select** the **POP2000** field as the **Population field**.

56. **Confirm** the ⊙ Kernel density option is set.

57. **Accept** the **default settings** for **Search radius**, and **Output cell size**.

58. **Save** the **Output raster** as **XX_KDens** (where **XX** is your state abbreviation) in your **student folder**.

59. **Click** [OK] to perform the density analysis. The density layer will be added to the map display.

60. **Open Layer Properties** for the **XX_Kdens** raster data layer.

61. **Double click** the symbol for the **lowest category** in the layer to open a **color palette**.

62.   *Click* the **No Color** option in the **color palette**.

63.   *Click* [OK] to apply the change and close **Layer Properties**.

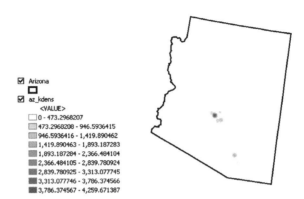

The map now displays the kernal density surface grid layer for your state. Notice the differences between the two density surfaces. As was the case in the previous exercise, the **kernel** density surface is smoother, while the **simple** density surface shows greater population away from population centers.

64.   *Save* 💾 the project.

Calculating Density of Features

Up until this point, you have performed density calculations dealing with population levels. You will now perform a density calculation to show the density of landscape features in a geographic area.

1.   *Insert* a new **data frame**.

2.   *Rename* the new data frame **Feature Density**.

3.   *Add* ⊕ the **XXXX_tgr#####cty00.shp** (where **XXXX** is your county abbreviation and ##### is your county TIGER code) county boundary data layer from the **C:\STARS\ExtToolsinSA\local** folder.

4.   *Add* ⊕ the county **XXXX_school.shp** data layer from the **C:\STARS\ExtToolsinSA\local** folder.

5.   To reproject this data into a measurable coordinate system, *double click* the **Feature Density** data frame to open **Data Frame Properties**.

6.   *Click* the **Coordinate System** tab.

Instead of selecting the coordinate system yourself as you did in the earlier activity, you will specify to use the coordinate system of another data layer using the **Import** option.

7. **Click** the | Import... | button to open the **Select a Data Source** dialog box.

8. **Navigate** to the **C:\STARS\ExtToolsinSA\local** folder.

9. **Select** the **XXXX_YYY_str_utm##_z##.shp** street shapefile.

10. **Click** | Add |.

   *Note: If, after following these steps your coordinate system is still "unknown", you will need to build it by selecting **Predefined/Projected Coordinate Systems** and selecting the projection parameters contained in the street shapefile name. This street shapefile can be found in the Surface\your_quad folder. For example, for shapefile **XXXX_YYY_str_utm27_z16n.shp**, the projection parameters are **UTM, NAD27, Zone 16N**.*

11. **Click** | OK | to apply the new coordinate system and close **Data Frame Properties**.

12. **Export** the **XXXX_tgr#####cty00** county boundary layer as a new shapefile named **YourCountyName_UTM.shp** in your **student folder**. Be sure to use **the same coordinate system as the data frame**.

13. When prompted, **add** the new data layer to the map.

14. **Remove** the **XXXX_tgr#####cty00** data layer.

15. **Export** the **XXXX_school** layer as a new shapefile named **YourCountyName_Schools_UTM.shp** in your **student folder**. Be sure to use **the same coordinate system as the data frame**.

16. When prompted, **add** the new data layer to the map.

17. **Remove** the **XXXX_school** data layer.

18. In the Table of Contents, *click* the **symbol** for the **YourCountyName_UTM** layer to open the **Symbol Selector**.

19. *Select* the | Hollow | symbol with a **2-point** outline.

20. *Click* [OK].

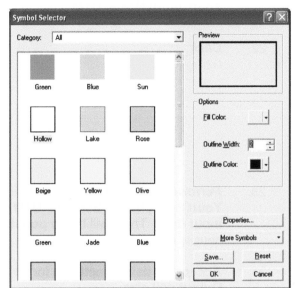

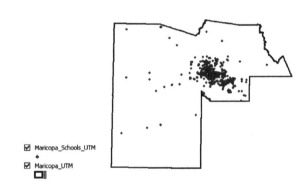

☑ Maricopa_Schools_UTM

☑ Maricopa_UTM

---

21. From the **Spatial Analyst** menu, *select* **Options** to *set* the **Extent** to the bounding coordinates of the **YourCountyName_UTM** projected county layer.

22. *Click* OK .

23. *Perform* a **kernal density** calculation with the **YourCountyName_Schools_UTM** school layer as the **Input data**. There will be **no population field** for this density calculation.

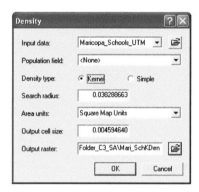

24. *Accept* the **default settings** for **Search radius**, and **Output cell size**.

25. *Save* the **Output raster** as **YYYY_SchKDen** (where **YYYY** is your county abbreviation) in your **student folder**.

26. *Click* OK .

27. *Uncheck* the **YourCountyName_Schools_UTM** layer.

28. *Alter* the **symbology** of this **density** layer as you did with the previous calculations.

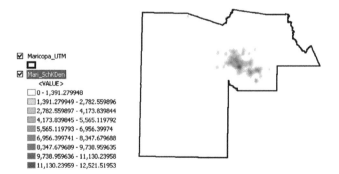

Creating the Map Layout:
You will now create a map layout to display the analysis results from this enrichment activity.

1. ***Switch*** to **Layout View** by clicking on the **Layout View**  button at the bottom of the Data View.

2. To change the page orientation of the layout page, ***select*** **Page and Print Setup...** from the **File** menu.

3. ***Set*** the ***Page Orientation*** to ***Landscape***.

4. Make sure that the ☑ Use <u>P</u>rinter Paper Settings option is checked.

5. ***Click*** ⬚ OK ⬚.

The data frames will be stacked on top of one another on the layout page.

6. ***Position*** the data frames by dragging them to new locations on the layout using your mouse.

7. ***Resize*** the data frames on the map layout page by clicking on the data frame boxes and using the sizing handles to click and drag.

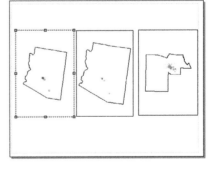

8. To place a title on the layout page, ***select*** **Title** from the **Insert** menu. When the default title is placed on the layout page, enter the title **Mapping Density – Enrichment Exercise**.

9. ***Press*** **Enter** on your keyboard to accept the title.

10. To change the font size, ***double click*** the **title** to open **Properties**.

11. ***Click*** the ⬚ Change Symbol... ⬚ button to open the **Symbol Selector**.

12. *Change* the font **size** to **24** with **bold** [B] **style**.

13. *Click* [OK] to close the **Symbol Selector**.

14. *Click* [OK] in **Properties** to apply the change to the title.

    Because each of the data frames has unique data layers, you must add a legend for each data frame.

15. *Click* on the **Simple Density** data frame.

16. *Select* **Legend...** from the **Insert** menu. The **Legend Wizard** dialog box will open.

17. Make sure that the **YourState** and **XX_Sdens** (where **XX** is your state abbreviation) layers are included as **Legend Items**.

25. When the legend is added to the map layout, *resize* it and *move* it to an appropriate place near the **Simple Density** data frame.

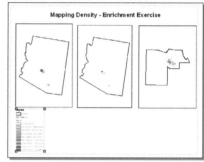

26. *Click* on the **Kernal Density** data frame.

27. *Insert* a **legend** for this data frame that includes the **YourState** and **XX_Kdens** data layers.

28. When the legend is added to the map layout, *resize* it and *move* it to an appropriate place near the **Kernal Density** data frame.

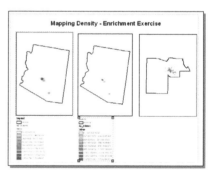

29. *Click* on the **Feature Density** data frame.

30. *Insert* a **legend** for this data frame that includes the **YourCounty_UTM** and **XXXX_SchKDen** data layers.

31. When the legend is added to the map layout, *resize* it and *move* it to an appropriate place near the **Feature Density** data frame.

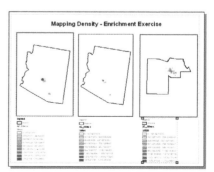

To make the first two (2) data frames have the same extent, you must make the scales the same for both data frames.

32. **Click** the **Simple Density** data frame.

33. **Click** in the **Map Scale** box to highlight the value in the box.

34. **Press** the **Ctrl** and **C** keys on the keyboard to copy the value.

35. **Click** the **Kernal Density** data frame.

The value in the **Map Scale** box probably differs from the one for the first data frame.

36. **Click** in the **Map Scale** box so that the value is highlighted.

37. **Press** the **Ctrl** and **V** keys on the keyboard to paste the scale from the first data frame into the **Map Scale** box for this data frame.

38. **Press** the **Enter** key on the keyboard to apply the new scale.

The size of the state boundary in the first two (2) data frames should be the same.

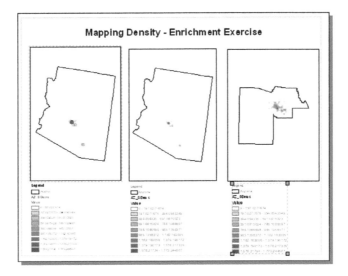

The scale of the third data frame should not be set the same as the first two because this data frame contains features at the county level and the features would be too small on the layout page.

You will now insert scale bars on the layout page. You could actually use one (1) scale bar for the first two (2) data frame because they are set at the same scale. You must add a separate scale bar for the **Feature Density** data frame, though.
Before you add any scale bar to the layout page, you must change the display units for the data frames. The display units are currently set as meters. Miles is a more appropriate unit of measure for a scale bar.

39. **Double click** the **Simple Density** data frame in the Table of Contents to open **Data Frame Properties**.

40. **Click** the **General** tab.

41. **Change** the **Display Units** to **Miles**.

42. **Click** [OK] to apply the change and close **Data Frame Properties**.

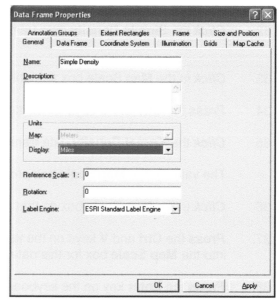

43. **Select** **Scale Bar...** from the **Insert** menu.

44. **Choose** a **scale bar style** in the **Scale Bar Selector** dialog box.

45. **Click** [OK] to add the scale bar to the layout page.

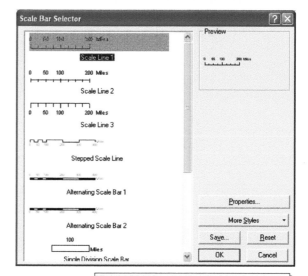

46. **Move** the **scale bar** to an appropriate place near the **Simple Density** data frame.

47. If necessary, **resize** the **scale bar**.

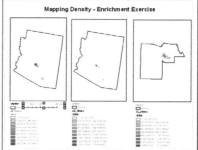

48. *Double click* the **Kernal Density** data frame in the Table of Contents to open **Data Frame Properties**.

49. *Click* the **General** tab.

50. *Change* the **Display Units** to **Miles**.

51. *Click* OK to apply the change and close **Data Frame Properties**.

52. *Select* **Scale Bar…** from the **Insert** menu.

53. *Choose* a **scale bar style** in the **Scale Bar Selector** dialog box.

54. *Click* OK to add the scale bar to the layout page.

55. *Move* the **scale bar** to an appropriate place near the **Kernal Density** data frame.

56. If necessary, *resize* the **scale bar**.

57. *Open* **Data Frame Properties** for the **Feature Density** data frame and *change* the **Display Units** to **miles**.

58. *Insert* a **scale bar** for the **Feature Density** data frame.

59. *Move* and *resize* the **scale bar** as needed.

You will now insert a north arrow on the layout page.

60. *Select* **North Arrow…** from the **Insert** menu.

61. *Select* a **north arrow style** in the **North Arrow Selector** dialog box.

62. *Click* OK to add the north arrow to the layout page.

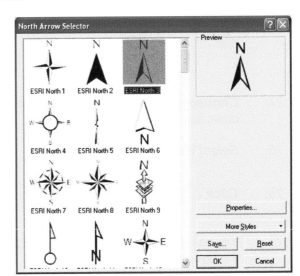

63. *Click* and *drag* the **north arrow** to an appropriate place on the layout page.

64. To place your **name** and the **date** on the layout page, *select* **Text** from the **Insert** menu. A text box will appear on the layout page (just as the title box did). It may be difficult to see the text if it is placed on another map element on the page.

65. *Type* your **name** and *press* **Enter** on your keyboard.

66. *Click* and *drag* the text box to a desired location on the map layout page.

67. *Double click* the **text box** to open **Properties** to add the date to the text box.

68. *Click* in the box next to your name and *press* **Enter** on your keyboard to go to the next line in the text box.

69. *Type* the **date**.

70. When you have finished, *click* OK .

71. To add a border to the layout page, *select* **Neatline...** from the **Insert** menu. The **Neatline** dialog box will appear.

72. *Choose* the Place inside margins option.

73. *Specify* a **gap** of **1 point** for the neatline.

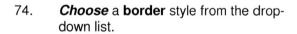

74. *Choose* a **border** style from the drop-down list.

75. *Select* **Hollow** as the **background**.

76. *Click* OK .

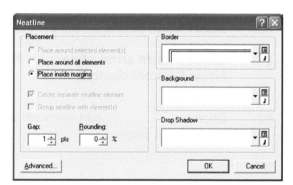

77. If necessary, *move* or *resize* any map elements on the layout page.

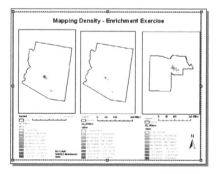

78. To export the map as an image file, *select Export Map...*from the *File* menu.

79. *Export* the layout to your **student folder** as **S3SAL2_Enrich_XX** (where **XX** is your initials) in **JPEG** format.

80. *Click* [Save] .

81. To print the layout, *select Print...* from the *File* menu.

82. In the **Print** dialog box, *click* [OK] to print the map layout page to your default printer.

83. *Save* 💾 this project.

84. Allow your instructor to see your work before you exit ArcMap.

85. *Select* **Exit** from the **File** menu or *click* the **Close** [X] button in the upper right corner of the ArcMap window to exit.

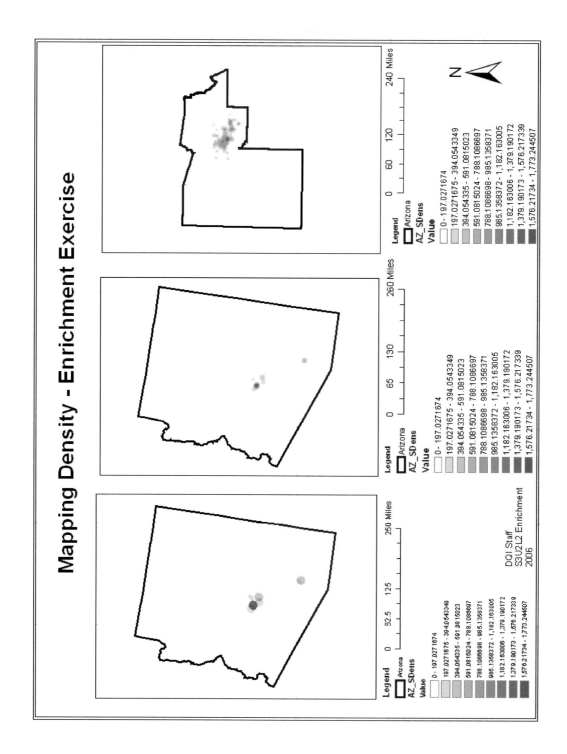

Lesson 2: Mapping Density Lesson Review

Key Terms
Use the lesson or index provided in the back of the book to define each of the following terms.

1. Simple density
2. kernel density
3. Point density

Global Concepts
Use the information from the lesson to answer the following questions. You may need to answer these on the back of this page or on your own paper.

4. What type of coordinate systems would you use for density measurements and why?

5. Explain why you might start with a simple density calculation.

6. In what circumstances is kernel density the more appropriate?

Let's Talk About It...
Answer the following question and share the responses with your instructor and classmates.

7. What fields of work might require regular usage of the density calculations and why?

Lesson 2

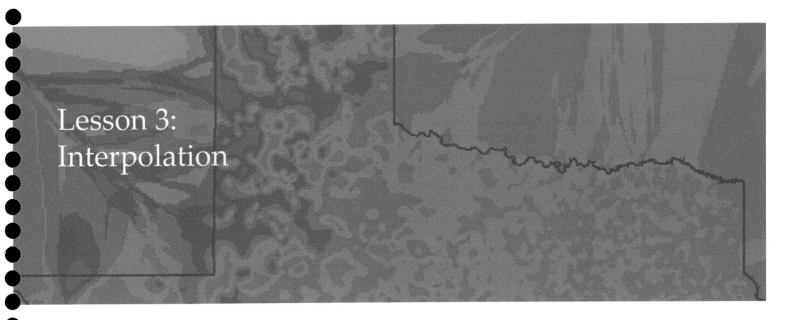

Lesson 3: Interpolation

Instead of undergoing a costly land survey in order to measure continuous data such as elevation, temperature, rainfall, etc. Spatial Analyst can be used to interpolate a surface, or create a surface, provided the user has values at specific locations in the study area.

Surface interpolation is a set of functions which can make predictions for all locations in a raster data set whether or not measurements have been taken at all locations. There are many different methods of interpolating data, each of which can provide accurate representation depending on data type and the assumptions made gathering the data.

When you think about it, nobody can measure every piece of surface data, and so there will always be a certain amount of estimation. With the use of interpolation, you are better able to 'fill in the blanks' in the data set, as well as work with information that has already been gathered, rather than making an attempt to collect more data points than can be handled.

There is no one 'right' interpolation method for all data sets or data types, and the primary way to select the appropriate interpolation is experience and comparison to other data sets. The difference between the interpolation methods is that each does a different mathematical equation to predict the results. The equations themselves are not so important for utilizing the software, but it is important to gain an understanding of what the different methods do to the data, which helps you to decide what method would be the best to try first on any given occasion.

Inverse Distance Weighted (IDW) reduces the contribution of distant points. The weight of each sample point is an inverse proportion to the distance, thus the name. The farther away a point, the less weight the point has in helping define the value at an unsampled location" *(Bolstad (2002:342)* This can be useful when trying to interpolate a data set where distance is one of the primary causes of the change in data. For example, if you are looking at temperatures across an area, unless there is a very specific reason, such as land formations or sea breezes, distance will be one of the primary reasons that temperatures change.

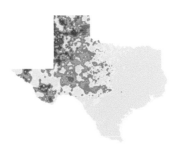

Spatial Technology And Remote Sensing

SPACESTARS

Laboratory for GIS/Remote Sensing Education

Spline interpolation is typically used when a set of points are collected along a line, such as a road, and then a smooth curved line is generated from those points to represent a continuous road on a map. The unknown locations between the known points are interpolated to fall along the curve created from the known points. Any time that the data follows a linear path, such as a travel route or movement patterns, spline interpolation might be indicated. An example of this might be migratory bird flight – where birds are sighted in several locations around the world at different points of time, and you want to connect their movements to find other possible locations to search for them.

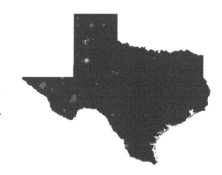

"The **Kriging** method assumes that the distance or direction between sample points reflects a spatial correlation that can be used to explain variation in the surface. Kriging fits a mathematical function to a specified number of points, or all points within a specified radius, to determine the output value for each location." *(Using ArcGIS Spatial Analyst, 2002)* When you use Kriging, you assume that the data points were taken for a specific reason and that you are working with that logic to better map out what happens. For example, if somebody is gathering information about topography and is marking heights at ridge points and

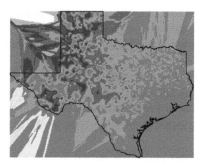

valleys and other areas where the changes occur, Kriging can be one of the best methods of interpolation.

Each of the methods is used extensively in geospatial analysis. No one method has been proven to be better or worse than another. It is dependent on the nature of the known sample points and study area. We will use all three methods now to generate an elevation surface of the state of Texas to see which one creates a more accurate representation of the true elevation of the state.

Lesson 3: Interpolation

In this lesson, you will use a point feature layer to build an elevation surface using interpolation tools. Interpolation tools are very valuable in surface analysis, as it makes it possible to estimate a continuous data surface without having to actually go out and collect data for every point on the surface of the study area. By taking known data point values, estimations can be made about the surface between the known points using different interpolation methods and a continuous surface can be created. Examples of this type of analysis include the creation of an elevation surface model generated from a set of specific elevation points; creation of temperature and precipitation maps from data collected at designated weather stations; and predictions of where certain mineral deposits might exist based on ground samples taken at designated testing locations.

1.   *Launch* **ArcMap** and start with **a new empty map**. *Save* the file as **S3SAL3_XX** in your student folder (where **XX** is your initials).

2.   *Add* ⊕ the **TX_Elev_pts** and **Texas_Bound** layers from the **C:\STARS\ExtToolsinSA** folder.

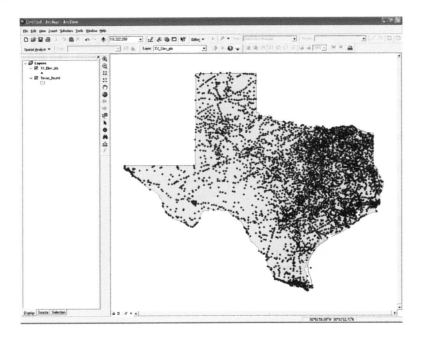

Because the map units for this data are currently in degrees minutes seconds (DMS), they must be changed to a measurable projected coordinate system in order to continue with the interpolation analysis.

3.   ***Right click the** Layers data frame.*

4.   ***Select** Properties*.

5.   ***Click** the **Coordinate System** tab.*

6.   ***Select** Predefined*.

7.   ***Select** Projected Coordinate Systems*.

8.   ***Select** State Systems*.

9.   ***Select** NAD 1927 Texas Statewide Mapping System*.

10.   *Click*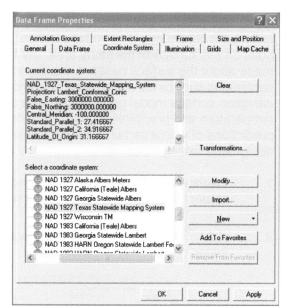

The map now changes to display into a measurable coordinate system, but the data layers themselves are not yet projected so that interpolations can be made.

11.   ***Right click** Tx_Elev_pts.*

12.   ***Select** Data*.

13.   ***Select** Export Data*.

14.   ***Use the same coordinates system as:***  .

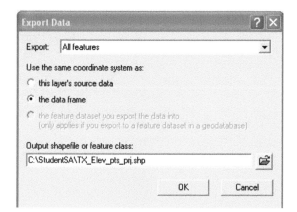

15.   ***Save** the **Output shapefile or feature class** in **your student folder** as **Tx_Elev_pts _prj**.*

      Click  .

16.   ***Add** the exported data to the map*.

17.   ***Remove** the non-projected layer **Tx_Elev_pts**.*

18.   ***Repeat** these steps to project the **Texas_Bound** layer and save the new projected layer as Texas_Bound_prj.*

19.   ***Remove** the non-projected layer **Texas_Bound***

Interpolation is the process of using discrete data points to create a continuous raster surface to describe an area. This is only useful in a situation where the entire surface would have these attributes. For example, rainfall, population density and elevation are all important to interpolate, but interpolating manhole covers or school locations would be meaningless because schools or manhole covers only exist at a single location.

There are three (3) types of interpolation methods included in **Spatial Analyst** functions: **Inverse Distance Weighted (IDW), Spline and Kriging**.

Each of the methods is used extensively in geospatial analysis. No one method has been proven to be better or worse than another. It is dependent on the nature of the known sample points and study area. We will use all three methods now to generate an elevation surface of the state of Texas to see which one creates a more accurate representation of the true elevation of the state.

"**IDW** interpolator estimates the value of unknown points using the distance and values to nearby known points. IDW reduces the contribution of distant points. The weight of each sample point is an inverse proportion to the distance, thus the name. The farther away a point, the less weight the point has in helping define the value at an unsampled location" *(Bolstad (2002:342)*

20.   *Rename* the data frame *"IDW"*.

21.   *Select* **Options...** from the Spatial Analyst ▼ drop-down menu.

22.   *Set* the **Analysis Mask** and **Extent** to the **Tx_Bound_prj** layer.

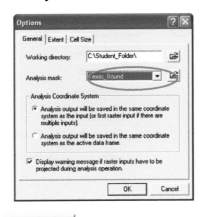 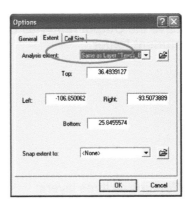

22.   *Click* OK .

23.   *Select* **Interpolate to Raster ▶ Inverse Distance Weighted...** from the Spatial Analyst ▼ drop-down menu.

24.   *Select* **TX_Elev_pts_prj** for the **Input points**.

25. **Set** Elevation as the **Z value** field.

The **Power** property listed on the IDW dialog box allows you to *"control the significance of known points on the interpolated values based on their distance from the output point."* The higher the Power number, the more emphasis placed on the nearest points. The resulting model will *"be less smooth and have more detail."* Alternatively a lower Power will give more influence to points farther away and result in a smoother surface. The most commonly used power value (the default) is 2.

"Inverse Distance Weighted (IDW) has two options: a fixed search radius type and a variable search radius type. With a **fixed** radius, the radius of the circle used to find input points is the same for each interpolated cell. By specifying a minimum count, you can ensure that within the fixed radius, at least a minimum number of input points will be used in the calculation of each interpolated cell. With a **variable** radius, the count represents the number of points used in calculating the value of the interpolated cell. This makes the search radius variable for each interpolated cell, depending on how far it has to stretch to reach the specified number of input points.

Specify a **maximum distance** to limit the potential size of the radius of the circle. If the number of points is not reached before the maximum distance of the radius is reached, fewer points will be used in the calculation of the interpolated point.

Use a barrier to limit the search for input sample points to the side of the barrier on which the interpolated cell sits— as in the case of a cliff or a ridge.

(Using ArcGIS Spatial Analyst, 2002)

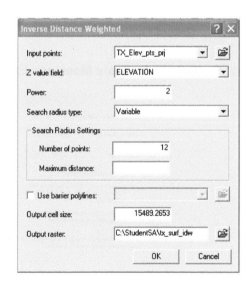

25. **Accept** all defaults for this process.

26. **Save** the **Output raster** as **tx_surf_idw** in your **student folder**.

27. **Rename** all three layers:

    TX_Elev_pts_prj as **Texas Elevation Points**.

    Texas_Bound_prj as **Texas Boundary**.

    tx_surf_idw as **Texas IDW**.

28. When the surface is added to the map, ***turn off*** the **Texas Elevation Points** and **Texas Boundary** data layers in the **Table of Contents**.

29. **Open** Layer Properties for the **Texas IDW** layer.

---

30.   ***Click*** the **Symbology** tab.

31.   ***Change*** the **color ramp** to the **Elevation #1** color ramp.

32.   ***Click*** OK .

The **IDW** interpolated surface will display using common elevation map colors in the map display.

The map displays an elevation surface showing the interpolated height of each cell in the grid based on the average values of the elevation points within its search radius.

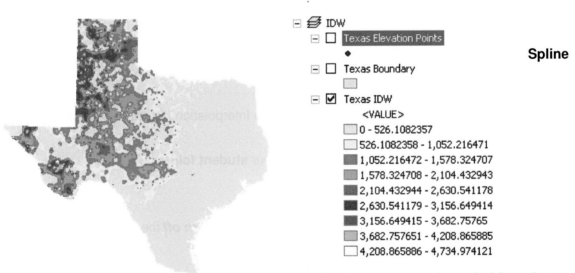

Spline

functions use a mathematical formula to smooth a curve created by "connecting the dots" of a known set of sample points. This type of interpolation is typically used when a set of points are collected along a line, such as a road, and then a smooth curved line is generated from those points to represent a continuous road on a map. The unknown locations between the known points are interpolated to fall along the curve created from the known points.

33.   ***Insert*** a new **data frame** into the **ArcMap** map document and ***rename*** it **Spline**.

34.   ***Drag*** the Texas Boundary and Texas Elevation Points layers into the new data frame.

35.   **Select Options...** from the Spatial Analyst ▼ drop-down menu.

36. *Set* the **Analysis Mask** and **Extent** to the **Texas Boundary Layer**.

37. *Select* **Interpolate to Raster ▶ Spline...** from the **Spatial Analyst ▼** drop-down menu.

   *"The **Regularized Spline** type ensures that you create a smooth surface and slope. The **Tension Spline** type tunes the stiffness of the surface according to the character of the modeled phenomenon.*

   *The **Number of points** option specifies the number of points used in the calculation of each interpolated point. The more input points you specify, the smoother the surface."*
   (ArcGIS Desktop Help)

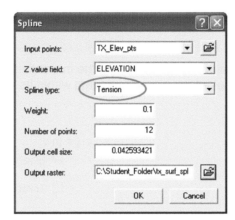

38. *Accept* the **default** values to perform a **Spline** Interpolation (change Spline type to **Tension** if it is not defaulted).

39. *Save* the **Output raster** as **tx_surf_spl** in your **student folder**.

40. *Click* **OK** .

41. When the surface model is added to the map display, *turn off* the **elevation points** and **Texas boundary** data layers in the **Table of Contents**.

42. *Open* **Layer Properties** and change the **color ramp** of the grid to ▼ (Elevation #1)

   just as you did with the **IDW** surface model.

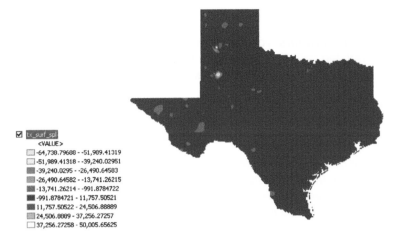

The map displays an elevation surface showing the interpolated height of each cell in the grid based on the average values of the elevation points in it's neighborhood, and a mathematical function that minimizes overall surface curvature. Though this interpolation results in a smoother surface than that resulting from the **IDW** interpolation, it is not an appropriate method to use in this case due to the small scale and relatively uniform relief of this dataset. The state of Texas covers a large area. By attempting to "string" along each point on a smooth curve at this scale, the variations in the relief of the landscape are smoothed, and therefore lost.

43.   ***Insert*** a new **data frame** into the **ArcMap** map document.

44.   ***Drag*** the **Texas boundary** and **elevation point layers** into the new data frame.

45.   ***Rename*** the new data frame **Kriging**.

46.   ***Reset*** the **Analysis Mask** and **Extent** to the **Texas Boundary** layer as you have done earlier in this lesson.

47.   ***Select*** **Interpolate to Raster ▶ Kriging** from the Spatial Analyst ▼ drop-down menu.

The **Kriging** method assumes that the distance or direction between sample points reflects a spatial correlation that can be used to explain variation in the surface. Kriging fits a mathematical function to a specified number of points, or all points within a specified radius, to determine the output value for each location." *(Using ArcGIS Spatial Analyst, 2002)* This method is most similar to IDW in that they both calculate a weighted average, though the Kriging method uses more complex geostatistical formulas to determine autocorrelation between known points in the sample.

*"**Ordinary Kriging** is the most general and widely used of the **Kriging** methods and is the default. It assumes the constant mean is unknown. This is a reasonable assumption unless there is some scientific reason to reject this assumption. Universal **Kriging** should only be used*

---

when you know there is a trend in your data and you can give a scientific justification to describe it.

*By using a **variable search radius**, you can specify the number of points to use in calculating the value of the interpolated cell. This makes the search radius variable for each interpolated cell, depending on how far it has to stretch to reach the specified number of input points.*

*Specifying a **maximum distance** limits the potential size of the radius of the circle. If the number of points is not reached before the maximum distance of the radius is reached, fewer points will be used in the calculation of the interpolated cell.*

With a fixed radius, the radius of the circle used to find input points is the same for each interpolated cell. The default radius is five times the cell size of the output grid. By specifying a minimum number of points, you can ensure that within the fixed radius, at least a minimum number of input points will be used in the calculation of each interpolated cell." (Using ArcGIS Spatial Analyst, 2002)

48.   ***Accept*** the **default** values to perform a **Kriging** Interpolation.

49.   ***Save*** the **Output raster** as **tx_surf_krg** in your **student folder**.

50.   ***Click*** `OK`.

51.   When the surface model is added to the map display, ***turn off*** the elevation points data layer in the **Table of Contents**.

52.   ***Rename*** the **tx_surf_krg** layer as **Texas Kirging**.

53.   ***Open*** **Layer Properties** and change the **color ramp** of the grid to

      (**Elevation #1**)

      just as you did previous surface models.

54.   ***Change*** the outline of the **Texas Boundary** layer to **No Fill, Outline Width: 2.00**

      Texas_Bound
      ▢   .

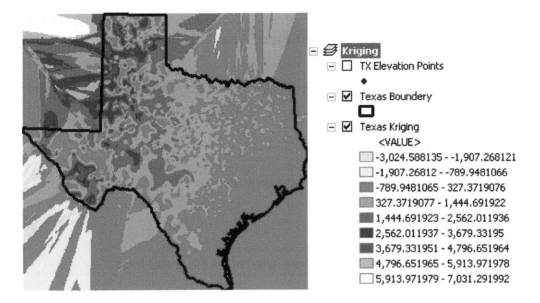

The map displays an elevation surface showing the interpolated height of each cell in the grid based on the statistical relationships among the elevation points. The map results in a similar surface to that resulting from the **IDW** interpolation, though the extent is not cropped to the state boundary. Notice, however, that since there are no elevation points outside of the boundary, that the surface features drawn there are distorted. They can, therefore, be ignored.

It appears from this analysis, that the **IDW** method of interpolation is best suited to generate this elevation surface model for the state of Texas. It stays true to the initial elevation point values, and "fills in the blanks" to create values for cells where there are no elevation point values that are consistent with the height and slope of the known values.

55.   ***Save*** the project.

Creating the Map Layout:

1.   *Switch* to **Layout View** by clicking on the **Layout View** ⬛ button at the bottom of the Data View.

2.   To change the page orientation of the layout page, *select* **Page and Print Setup...** from the **File** menu.

3.   *Set* the **Page Orientation** to **Landscape**.

4.   Make sure that the ☑ Use Printer Paper Settings option is checked.

5.   *Click*   OK   .

     The data frames will be stacked on top of one another on the layout page.

6.   *Position* the data frames by dragging them to new locations on the layout using your mouse.

7.   *Resize* the data frames on the map layout page by clicking on the data frame boxes and using the sizing handles to click and drag.

8.   To place a title on the layout page, *select* **Title** from the **Insert** menu. When the default title is placed on the layout page, enter the title **Comparing Interpolation Methods**.

9.   *Press* **Enter** on your keyboard to accept the title.

10.   *Change* the **title** font **size** to **24** with **bold** **B** style.

11.   **Insert** a **legend** for each of the data frames.

12.   Make each data frame have the same extent by *using* the **map scale** for each data frame.

13.   *Insert* a **scale bar** on the layout page using appropriate display units.

14.   *Insert* a **north arrow** on the layout page.

15.   *Insert* your **name** and the **date** on the layout page.

16.   *Insert* text to appropriately label each data frame on the layout page.

17.   *Insert* a **neatline** on the layout page.

18.   *Export* the map **S3SAL3_XX** (where **XX** is your initials) in **JPEG** format to your **student folder**.

19. *Print* the layout.

20. *Save* 🖫 the project.

21. Allow your instructor to see your work before you exit ArcMap.

22. *Select* **Exit** from the **File** menu or *click* the **Close** ☒ button in the upper right corner of the ArcMap window to exit.

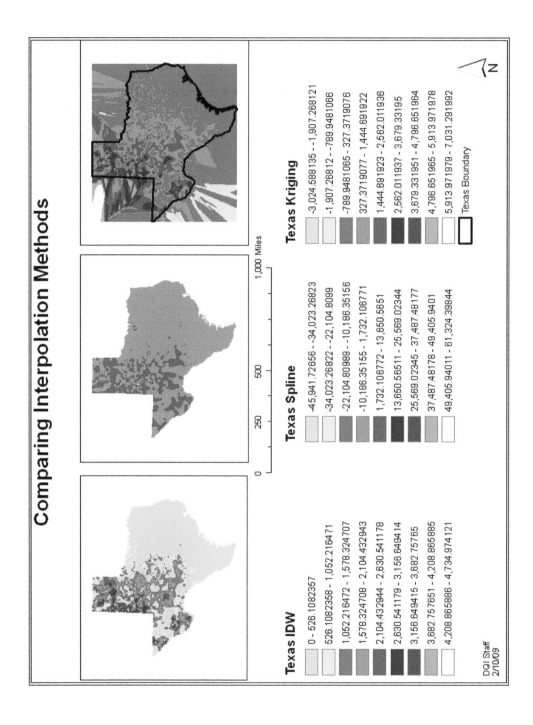

Lesson 3: Enrichment Activity

In the regular lesson activity, you interpolated a surface of the state of Texas from a series of elevation points. By using different interpolation techniques, you produced different results. In this enrichment activity, you will apply the same skills and techniques to your state data. (If you are a Texas school or would like additional practice with interpolation after completing your state's analysis, use the **CA_Elevation.shp** and layer located in the **C:\STARS\ExtToolsinSA\local** folder.)

1. *Launch* **ArcMap** with a **new blank map**.

2. If ArcMap is still in **Layout View**, *click* to change to **Data View** .

3. *Save* the project as **S3SAL3_Enrich_XX.mxd** (where XX is your initials) in your **student folder**

4. *Add* the **ZZ_elev_pts.shp** data layer (where **ZZ** is your state abbreviation) located in the **C:\STARS\ExtToolsinSA\local** folder.

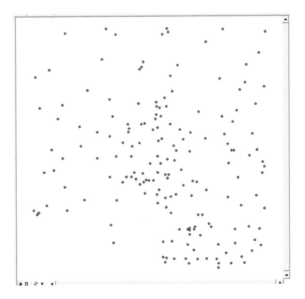

Notice that the data is displayed in degrees minutes seconds (Geographic Coordinate System).

3.   ***Add*** 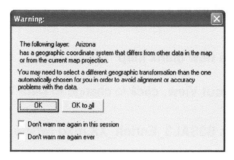 the **YourState.shp** layer that you created in Lesson 2 Enrichment Activity from your **student folder.** (If you are using the California data, use the same method from Lesson 2 Enrichment Activity to select and export the California state boundary.)

When the state boundary data layer is added, the following warning dialog box will appear to inform you that the data layer that you are adding has a coordinate system that differs from the data that is already in the map. (Remember that you exported your state boundary using the **USA Contiguous Albers Equal Area Conic** coordinate system.)

Warning:

The following layer: Arizona
has a geographic coordinate system that differs from other data in the map
or from the current map projection.

You may need to select a different geographic transformation than the one
automatically chosen for you in order to avoid alignment or accuracy
problems with the data.

[OK]   OK to all

☐ Don't warn me again in this session
☐ Don't warn me again ever

4.   ***Click*** [OK] and ArcMap will reproject the data "on-the-fly" so that it can be displayed in the map.

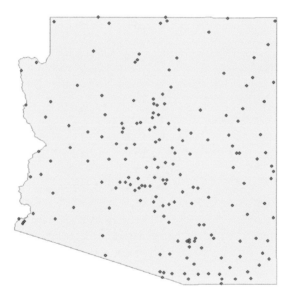

4.   In three separate **data frames** of this project use the three different interpolation methods to create a surface model of your state. Refer to the instructions for the regular lesson activity for **step-by-step instructions** on performing this analysis.

5.   ***Save*** the **output rasters** as **XX_IDW**, **XX_Spline** and **XX_Kriging** (where **XX** is your state abbreviation) as they apply. It may take several minutes for these to process.

6. *Edit* the **Symbology** for each interpolation map to use the color ramp and line weights as used in the earlier lesson on interpolation.

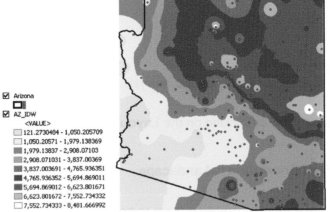

Arizona

AZ_IDW
<VALUE>
- 121.2730484 - 1,050.205709
- 1,050.20571 - 1,979.138369
- 1,979.13837 - 2,908.07103
- 2,908.071031 - 3,837.00369
- 3,837.003691 - 4,765.936351
- 4,765.936352 - 5,694.869011
- 5,694.869012 - 6,623.801671
- 6,623.801672 - 7,552.734332
- 7,552.734333 - 8,481.666992

Inverse Distance Weighted Interpolation of Arizona

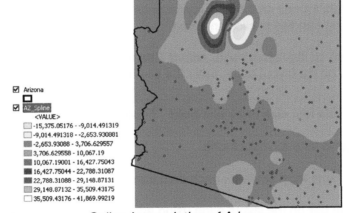

Arizona

AZ_Spline
<VALUE>
- -15,375.05176 - -9,014.491319
- -9,014.491318 - -2,653.930881
- -2,653.93088 - 3,706.629557
- 3,706.629558 - 10,067.19
- 10,067.19001 - 16,427.75043
- 16,427.75044 - 22,788.31087
- 22,788.31088 - 29,148.87131
- 29,148.87132 - 35,509.43175
- 35,509.43176 - 41,869.99219

Spline Interpolation of Arizona
Note: colors apply only to the range of elevation within each state.

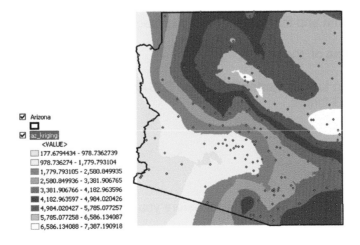

Arizona

az_kriging
<VALUE>
- 177.6794434 - 978.7362739
- 978.736274 - 1,779.793104
- 1,779.793105 - 2,580.849935
- 2,580.849936 - 3,381.906765
- 3,381.906766 - 4,182.963596
- 4,182.963597 - 4,984.020426
- 4,984.020427 - 5,785.077257
- 5,785.077258 - 6,586.134087
- 6,586.134088 - 7,387.190918

Kriging Interpolation of Arizona

7.   **Review** each of the interpolated maps and determine which method generated the most accurate and realistic elevation model of your state.

8.   **Save** 💾 the project.

Creating the Map Layout:
You will now create a map layout to display the analysis results from this enrichment activity.

1.   **Switch** to **Layout View** by clicking on the **Layout View** button at the bottom of the Data View.

2.   **Change** the **Page Orientation** to **Landscape**.

3.   **Position** the data frames by dragging them to new locations on the layout using your mouse.

4.   **Resize** the data frames on the map layout page by clicking on the data frame boxes and using the sizing handles to click and drag.

5.   To place a title on the layout page, **select** Title from the **Insert** menu. When the default title is placed on the layout page, enter the title **Interpolation Methods – Enrichment Exercise**.

6.   **Change** the **title** font **size** to **24** with **bold** **B** style.

7.   **Insert** a **legend** for each data frame.

8.   **Specify** the same **map scale** for each data frame.

9.   **Insert** a **scale bar** on the layout page using appropriate display units.

10.   **Insert** a **north arrow** on the layout page.

11.   **Insert** your **name** and the **date** on the layout page.

12.   **Insert** a **neatline** on the layout page.

13.   **Export** the map to your **student folder** as **S3SAL3_Enrich_XX** (where **XX** is your initials) in **JPEG** format.

14.   **Print** the layout.

15.   **Save** 💾 the project.

16.   Allow your instructor to see your work before you exit ArcMap.

17. **_Select_ Exit** from the **File** menu or **_click_** the **Close** ☒ button in the upper right corner of the ArcMap window to exit.

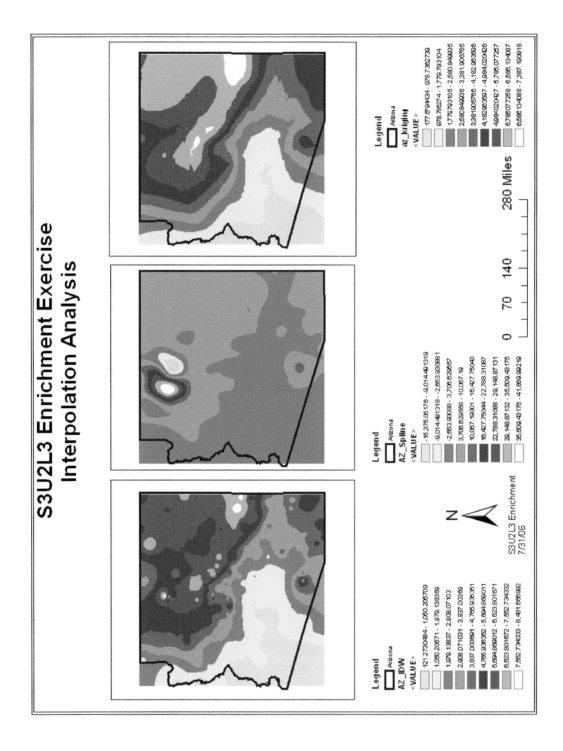

Lesson 3: Interpolation Lesson Review

Key Terms
Use the lesson or index provided in the back of the book to define each of the following terms.

      1. Interpolation
      2. Inverse Distance Weighted
      3. Spline
      4. Kreiging

Global Concepts
Use the information from the lesson to answer the following questions. You may need to answer these on the back of this page or on your own paper.

    5.   Is there a single correct form of interpolation? Why or why not?

    6.   If you performed an interpolation for precipitation, what might you compare it to in order to see which one was closest to reality and why?.

    7.   If you were studying migration patterns or other linear movement patterns, which form of interpolation would you start with?

Let's Talk About It...
Answer the following question and share the responses with your instructor and classmates.

    8.   What fields of work might require regular usage of the interpolation and why?

Lesson 4:
Analyzing Surfaces

Spatial Analyst allows you to analyze the surface of a study area in order to determine how the terrain will affect a particular problem or situation. The surface Analysis tools available to users are: Aspect, Contour, Curvature, Cut/Fill, Hillshade, Slope, and View Shed.

Surface analysis allows you to identify patterns in an original data layer that could not have been identified without the analysis. Using Spatial Analyst with elevation data allows you to create elevation contours, determine the slope of the landscape in the study area, identify the aspect of the landscape, also, create a hillshade that shows terrain relief, and evaluate visibility using viewshed analysis and estimate change in areas and volumes using cut/fill analysis.

The contours function can be used for any dataset where you want to show variations across the surface, whether it is 'real' surface such as the ground(elevation contours) or a theoretical surface such as rainfall levels, pollution levels, atmospheric pressure, or even frequency of emergency calls. The contours are created using polylines to show how the values change over the surface. The lines connect all known points at a particular 'elevation' and suggest where elevation changes rapidly and slowly – the closer together the lines are, the more rapidly the elevation is changing, while the further apart the lines are, the more slowly the elevation is changing.

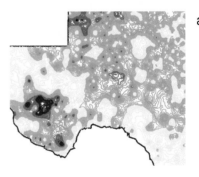
a

The slope function can be used on any set of contours. Slope allows you to calculate the elevation change between any raster cell and the surrounding raster cells. The lower the slope, the flatter the terrain; the higher the slope value, the steeper the terrain. The higher slope areas are most likely found along ravines and ridges, while lower slope areas are more open spaces. The slope calculation is very valuable for considerations such as land usage. After all, it is considerably more difficult to build a house on a steep slope than on a level piece of ground. Another consideration is that water always flows downhill, and the force of gravity ends up pulling it down the steepest slope available. This means that by knowing the slope, you can also predict

Spatial Technology And Remote Sensing
STARS
SPACESTARS
Laboratory for GIS/Remote Sensing Education

things like areas in the greatest danger of flooding or the most likely locations of streams, depending on your area of interest.

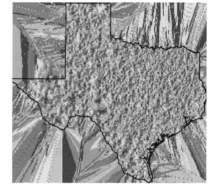

Aspect refers to the downslope direction of a landscape. It is sometimes referred to as slope direction. Aspect is measured clockwise in degrees with 0° being due north. Examples of how this analysis is used include finding north-facing slopes for ski routes, calculating solar illumination, evaluating meltwater effects from southerly slopes and identifying flat land for an emergency airplane landing. The values derived for aspect indicate the direction of downslope for each grid cell to its neighbor. The direction is classified into 10 intervals on the basis of a 360 degree compass circle. From a surface grid analysis such as this, all hills facing north can be identified as potential ski slopes, or south facing slopes can identified for particular agricultural cultivation.

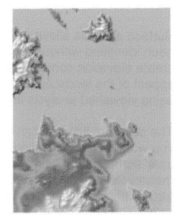

The Spatial Analyst Hillshade function allows you to set a light source position to produce a surface that enhances the appearance of its relief. This enhancement creates a more realistic image of the landscape which may assist in the planning and development of a region. Using the Hillshade function also allows you to determine which areas will be shaded at particular times of the day. This type of analysis is helpful for agricultural planning, as well as site planning for residential, commercial and recreational development. These displays may assist planning efforts when communicating the impacts and effects that proposed projects may have on conditions of the area. Another important aspect is that by utilizing hillshade analysis, it becomes much easier to show how much of a factor elevation is over a wide surface. People are used to seeing areas with light on them, so by running hillshade, you make the elevation information more familiar and easily understood to people, especially when it is layered with color-coded elevation data.

Viewshed identifies places in a study area that can be seen from a particular observation point. This analysis is useful for applications such as law enforcement planning when determining point locations for a stake out; or the siting of new cellular telephone towers along a transmission corridor adjacent to a residential community. Effective uses include the design of interpretive signage informing park visitors at this observation point what peaks, valleys and other significant landmarks can be viewed from this location.

Analyzing cut/fill is another process that combines several aspects of spatial analysis. It helps to see how the contour and character of the earth's surface are changing. Usually, these changes are the result of gradual forces, such as erosion or development. However, sudden climatic or seismic events, such as volcanic eruptions or tsunamis, can cause tremendous short term change to the surface of a geographic area. The DEMs are of Mount St. Helens before the devastating 1980 eruption and after. Areas with a gain of surface volume are displayed in red, while areas with a net loss in surface volume are displayed in blue. What does this tell you about the Mount St. Helens area and how can an analysis like this be useful to scientists and planners studying the impacts of natural disasters such as this?

Lesson 4: Analyzing Surfaces

One of the great advantages of using the **Spatial Analyst** extension is its ability to perform varied surface analyses. Surface analysis allows you to identify patterns in an original data layer that could not have been identified without the analysis. Using **Spatial Analyst** with elevation data allows you to create elevation contours, determine the slope of the landscape in the study area, identify the aspect of the landscape, create a hillshade that shows terrain relief, evaluate visibility using viewshed analysis and estimate change in areas and volumes using cut/fill analysis. You will explore these functions in this lesson.

Creating Contours

The **Contours** function in the **Spatial Analyst/Surface Analysis** tools allows you to create contours from a data layer. **Contours** are polylines that connect points of equal value, such as elevation, temperature, precipitation, pollution, or atmospheric pressure. The distribution of a polyline shows how values change across a surface. Where there is little change in a value, the polylines are spaced farther apart. Where the values rise and fall rapidly, the polylines are closer together. *(ArcGIS Desktop Help)*

In this exercise, you will generate a set of elevation contour lines from the Texas grid surface model that you created from the set of statewide elevation feature points.

1. *Launch* ArcMap with a **new map document**.

2. *Save* 💾 the map document as **S3SAL4_XX.mxd** (where **XX** is your initials) in your **student folder**.

3. *Add* ✛ the **tx_surf_idw** grid located in your **student folder**.

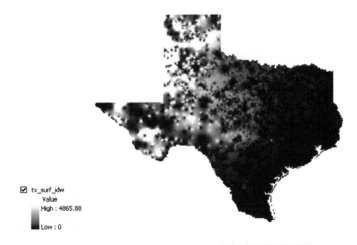

☑ tx_surf_idw
  Value
  High : 4865.88

  Low : 0

4. *Select* **Surface Analysis/Contour...** from the Spatial Analyst ▼ drop-down menu.

The **Contour interval** specifies the distance between contour polylines. The wider the range in elevation, the higher the contour interval.

The **Base contour** is the value from which to begin generating contours. Contours are generated above and below this value as needed to cover the entire value range of the grid.

The **Z Factor** is the number of ground x,y units in one surface z unit. The Input surface values are multiplied by the specified z-factor to adjust the Input surface z units to another measurement unit. So, for instance, if your x and y units are in meters, and your z units are in feet, you would specify a z-factor of 0.3048, as there are 0.3048 meters in one foot. For this exercise, we will maintain the z units as feet, so the default value of 1 is accepted.

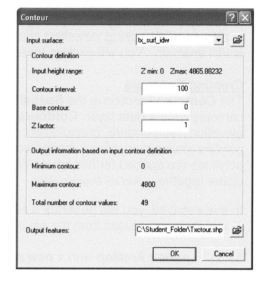

5.   ***Accept*** all other **default** values.

6.   ***Save*** the **Output features** (contours) as **TXctour.shp** in your **student folder**.

7.   ***Click*** OK .

8.   ***Open*** **Layer Properties** for the **tx_surf_idw** data layer.

9.   ***Change*** the **symbology** so that the elevation data is displayed with the **Elevation #1** color ramp ⬛⬛⬛⬛⬛ ▾ .

10.   ***Add*** ✛ the **Texas_Bound** layer from the **C:\STARS\ExtToolsinSA** folder.

11.   ***Change*** the **symbology** to **Hollow** with an **Outline Width** of **2.00** ☐ .

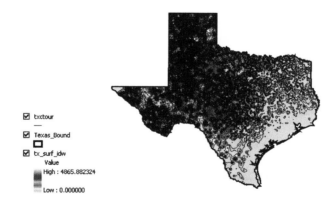

12.   ***Zoom in*** 🔍 within the state boundary to view the detail of the contours.

Extended Tools in Surface Analysis, version 7

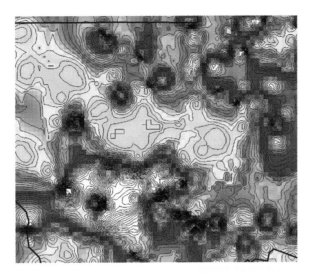

While the elevation surface grid shows the height of each cell by assigning it a specific color within an elevation range, e.g., white for the highest elevation cells, the contours enhance the model by showing not only the height, but relative steepness or flatness, based on the distance between contours. The closer the contours, the steeper the terrain is.

Each contour line represents a different height, connecting points or cells of common elevation.

13.   ***Zoom out*** 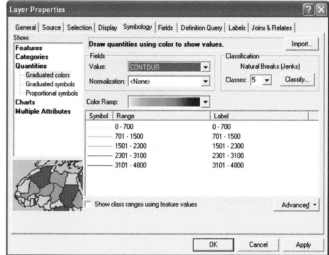 to the full extent of the map.

14.   ***Turn off*** the **tx_surf_idw** layer.

15.   ***Open*** the **Layer Properties** for the **TXctour** data layer.

16.   ***Change*** to show the **symbology** of **TXctour** as **Quantities/Graduated Colors**.

17.   ***Select CONTOUR*** as the **Value field** to view the variation in these contour values.

18.   ***Select*** the **Yellow to Green to Dark Blue color ramp**
.

19.   ***Click*** OK to apply the change and close **Layer Properties**.

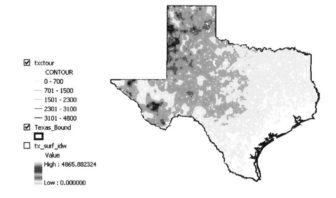

20.   ***Zoom in*** within the state boundary to view the detail of the contours.

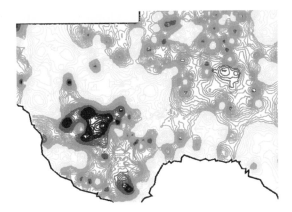

The variation in the elevation is now visible based on the classification and coloring of the contours into intervals of varying elevation.

☑ TXctour
    CONTOUR
    — 0 - 700
    — 701 - 1500
    — 1501 - 2300
    — 2301 - 3100
    — 3101 - 4800

21.   *Rename* the data frame **Contours**.

22.   *Save* the project.

Generating Slope

In Lesson 1, you generated a slope surface to perform a cost weighted distance and allocation function. In this exercise, you will generate a slope surface of a section of York County, Pennsylvania.

1.   *Insert* a new **data frame** in the project. *Rename* it **Slope**.

2.   *Add* ⬇ the **yorkpa_dem** layer from the **C:\STARS\ExtToolsinSA** folder.

3.   *Select* **Surface Analysis ▶ Slope...** from the Spatial Analyst ▼ drop-down menu.

4.   *Confirm* the **yorkpa_dem** data layer is set as the **Input surface**.

5.   *Accept* **Degree,** the default **Z Factor 1**, and **Output cell size.**

6.   ***Specify*** to save the **Output raster** as **York_slope** in your **student folder**.

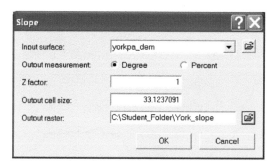

7.   ***Click*** OK .

Slope is calculated as the maximum rate of elevation change over distance between each raster cell and its neighbors. The lower the slope, the flatter the terrain; the higher the slope value, the steeper the terrain. In this case, much of this section of York County is at or below a 10% slope, though there are a number of locations where the slope reaches as high as almost 38%. These locations are most likely found along ravines and ridges.

8.   ***Uncheck*** the **yorkpa_dem** raster data layer in the Table of Contents.

9.   ***Save*** the project

Identifying Aspect

Aspect refers to the downslope direction of a landscape. It is sometimes referred to as slope direction. Aspect is measured clockwise in degrees with 0° being due north. Examples of how this analysis is used include finding north-facing slopes for ski routes, calculating solar illumination, evaluating meltwater effects from southerly slopes and identifying flat land for an emergency airplane landing.

1. *Insert* a new **data frame** in the project. *Rename* it **Aspect**.

2. *Drag* the **tx_surf_idw** and **Texas_Bound** layers located in the **Contours** data frame into this new data frame.

3. *Minimize* the **Contours** and **Slope** data frames by *clicking* on the **minus signs** ⊟ to

    the left of the data frame so that they become **plus signs**.

4. *Select* **Surface Analysis ▶ Aspect...** from the Spatial Analyst ▼ drop-down menu.

5. *Confirm* the **tx_surf_idw** Texas elevation grid is set as the **Input surface**.

6. *Accept* the default **Output cell size**.

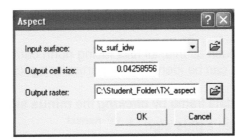

7. *Save* the **Output raster** as **TX_aspect** in your **student folder**.

8. *Click* OK .

The aspect derived from the statewide elevation grid of Texas will process and display.

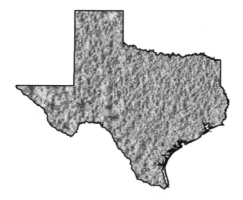

The values derived for aspect indicate the direction of downslope for each grid cell to its neighbor. The direction is classified into 10 intervals on the basis of a 360-degree compass circle.

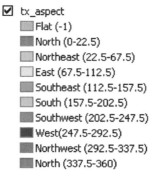

From a surface grid analysis such as this, all hills facing north can be identified as potential ski slopes, or south-facing slopes can be identified for particular agricultural cultivation.

9.   ***Collapse*** the **Aspect** data frame by ***clicking*** the **minus sign** □ to the left of the data frame so that it becomes a **plus sign**. ⊞ 🗇 **Aspect**

10.   If you turned on the **tx_surf_idw** data layer at any point in this analysis, ***uncheck*** it now.

11.   ***Save*** the project.

<u>**Displaying Hillshade**</u>
The **Spatial Analyst Hillshade** function allows you to set a light source position to produce a surface that enhances the appearance of its relief. This enhancement creates a more realistic image of the landscape which may assist in the planning and development of a region.

Using the **Hillshade** function also allows you to determine which areas will be shaded at particular times of the day. This type of analysis is helpful for agricultural planning, as well as site planning for residential, commercial and recreational development.

1.   ***Insert*** a new **data frame** in the project.

2.   ***Rename*** it **Hillshade**.

3.   ***Add*** ⬇ the **sfn_dem** grid located in the **C:\STARS\ExtToolsinSA** folder. This is a digital elevation model (DEM) of the North San Francisco quadrangle.

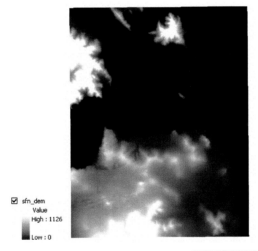

☑ sfn_dem
   Value
   High : 1126

   Low : 0

4.   ***Select* Surface Analysis ▶ Hillshade...** from the ⌄ Spatial Analyst ⌄ drop-down menu.

5.   ***Accept*** the **default Output cell size**.

6.   ***Specify*** to save the **Output raster** as **SFN_hillshade** in your **student folder**.

7.   ***Accept*** the **default Azimuth** and **Altitude** values.

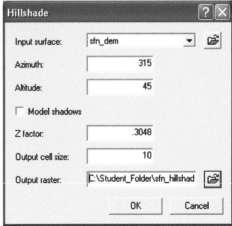

Azimuth refers to the angle of the sun measured clockwise in degrees with 0° being due north. The default azimuth value is 315° or northwest.

Altitude is the angle of the sun above the horizon and is measured in degrees from 0 to 90 with 0 being at the horizon and 90 being directly overhead. The default altitude value is 45°.

8.　　　***Enter*** a value of **0.3048** for **Z factor**.

Z factor allows you to make the z (elevation) values in the same units as the x,y coordinates. In this case, the elevation units are in feet, and the x,y map units are in meters because it is a UTM projected DEM layer. For this reason, entering a value of 0.3048 converts the resulting units to meters. The default z-value is 1.

9.　　　***Click*** ⬚ OK ⬚. The hillshade appears in the map display.

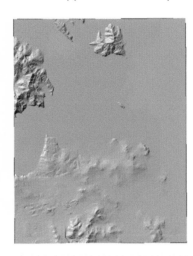

10. ***Drag*** the **sfn_dem** elevation data layer above the **hillshade** layer in the **Table of Contents.**

11. ***Open*** **Layer Properties** for the **sfn_dem** layer.

12. ***Click*** the **Symbology** tab and ***change*** the **color ramp** to

 (**Elevation #1**).

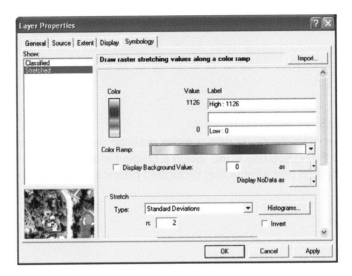

You will now add a three-dimensional appearance to this map display by making the DEM layer transparent.

13. To access the **Effects** toolbar, ***select*** **Toolbars ▶ Effects** from the **View** menu.

14. ***Confirm*** that the elevation grid **sfn_dem** is specified as the **layer Effects** toolbar

15. ***Click*** the **Adjust Transparency** tool . ***Adjust*** the transparency slider bar to approximately 60% so that the elevation colors of the elevation grid appear to drape over the hillshade relief.

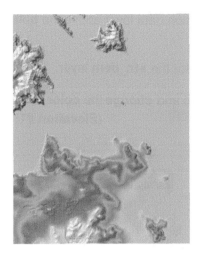

The appearance of this map display is now a more realistic depiction of the elevation of the north quad of the San Francisco bay area. Such a display may assist planning efforts when communicating the impacts and effects that proposed projects may have on conditions of the area.

16.   *Save* this project.

Analyzing Viewshed
Viewshed identifies places in a study area that can be seen from a particular observation point. This analysis is useful for applications such as law enforcement planning when determining point locations for a stake out; or the siting of new cellular telephone towers along a transmission corridor adjacent to a residential community.

1.   *Insert* a new **data frame** in the project.

2.   *Rename* it **Viewshed**.

3.   *Collapse* the ⊞ 📑 Hillshade data frame.

4.   *Add* ✛ the **Leconte_DEM** grid and the **Leconte_obs.shp** data layer located in the **C:\STARS\ExtToolsinSA** folder. This is a digital elevation model (DEM) of the Mount Leconte quadrangle in the Great Smoky Mountains National Forest and a point representing an observation area on the mountain.

☑ leconte_dem
Value
High : 2004.3
Low : 398.6

5.   *Select* **Surface Analysis ▶ Viewshed...** from the Spatial Analyst ▼ drop-down menu.

6.   *Confirm* that **leconte_dem** is set as the **Input surface** and **LeConte_obs** is set as the **Observer points**.

7.   *Accept* **1** as the **Z factor** since the x,y and z values are all in meters.

8.   *Accept* the **default Output cell size**.

9.   *Save* the **Output raster** as **LC_Viewshed** in your **student folder**.

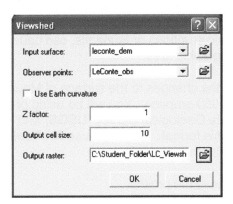

10.   *Click* OK.

After processing, the viewshed analysis appears in the map display.

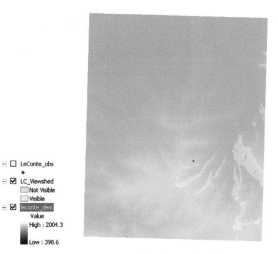

By adding additional geographic reference points, this map can be used to determine viewing corridors from an observation point. Effective uses include the design of interpretive signage informing park visitors at this observation point what peaks, valleys and other significant landmarks can be viewed from this location.

11.   *Save* this project.

Analyzing Cut/Fill

The contour and character of the earth's surface are always changing. Usually these changes are the result of gradual forces, such as erosion or development. However, sudden climatic or seismic events, such as volcanic eruptions or tsunamis, can cause tremendous short term change to the surface of a geographic area.

In this activity you will analyze the changes to the surface of Mount St. Helens, which occurred as a result of the devastating 1980 eruption. You will be using pre- and post-eruption data that is distributed by the United State Geological Survey (**USGS**) and is downloadable from the USGS website in **DEM** (.dem) file format. You must use **ArcToolbox** to convert these files to raster grids before using them for this analysis.

1.   *Insert* a new **data frame** in the project.

2.   *Rename* it **Cut/Fill**.

3.   *Collapse* the ⊞ 🗂 Viewshed data frame.

4.   *Click* the **Show/Hide ArcToolbox Window** 📦 button to display the **ArcToolbox Wndow** in **ArcMap**.

5.   ***Double click*** the **Conversion Tools** toolbox to display the toolsets contained in it.

6.   ***Double click*** the **To Raster** toolset to display the tools contained in it.

7.   ***Double click*** the **DEM to Raster** conversion tool. The **DEM to Raster** dialog box appears.

8.   ***Click*** the **Browse** 📂 button to the right of the **Input USGS DEM file** box and ***navigate*** to **C:\STARS\ExtToolsinSA**.

9.   ***Select*** the **oldmtsthelens.dem** file

10.   ***Click***   Open   .

11.   In the **Output raster** box, ***save*** the new grid as **Old_MSH** in your **student folder**.

12.   ***Accept* FLOAT** as the **Output data type**, and **1** as the **Z factor** since the x,y and z values are all in meters.

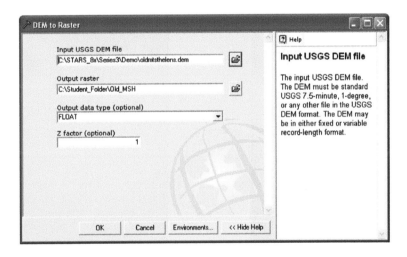

13.   ***Click***   OK   .

14.   When processing is complete for the old Mount St. Helens DEM, ***click*** | Close | in the **DEM to Raster** dialog box.

15.   ***Repeat*** the same process for the **newmtsthelens.dem** file located in the **C:\STARS\ExtToolsinSA** folder.

16.   ***Save*** the output raster as **New_MSH** in your **student folder**.

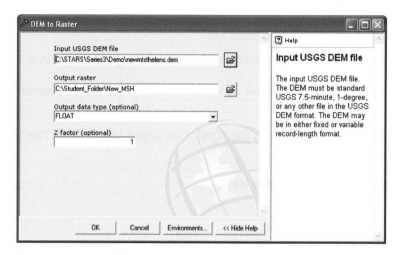

17. When you have finished converting these files, *close* **ArcToolbox** ✕.

18. *Toggle* the layers **on and off** as necessary in the **Table of Contents** to view both layers.

Old_MSH: Pre-eruption   **New_MSH: Post-eruption**

To determine areas that have changed as a result of the volcanic eruption in 1980, you will use the **Cut/Fill** option.

19. *Select* **Surface Analysis ▶ Cut/Fill** from the Spatial Analyst ▼ drop-down menu.

20. In the **Cut/Fill** dialog box, *select* **Old_MSH** as the **Before surface** and **New_MSH** as the **After surface**.

21. *Accept* the default **Z factor** and **Output cell size**.

22. *Save* the **Output raster** as **MSH_CutFill** in your **student folder**.

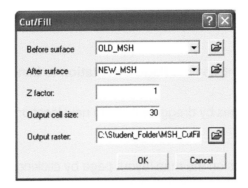

23. **Click** OK . The **Cut/Fill** analysis grid displays in the map.

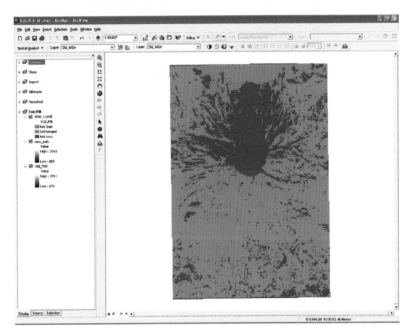

Areas with a gain of surface volume are displayed in red, while areas with a net loss in surface volume are displayed in blue. What does this tell you about the Mount St. Helens area and how can an analysis like this be useful to scientists and planners studying the impacts of natural disasters such as this?

24. **Uncheck** the **new_msh** and **old_msh** layers in the Table of Contents.

25. **Save** this project.

<u>Creating the Map Layout:</u>

1. *Switch* to **Layout View** by clicking on the **Layout View** button at the bottom of the Data View.

2. *Change* the page orientation the **Page Orientation** to **Landscape**.

3. *Position* the data frames by dragging them to new locations on the layout using your mouse.

4. *Resize* the data frames on the map layout page by clicking on the data frame boxes and using the sizing handles to click and drag.

5. *Insert* the following **title** on the layout page in **24-point bold** type:

    Surface Analysis Methods

6. **Insert** a **legend** for each data frame on the map layout page.

7. *Insert* a **scale bar** for each data frame on the layout page using appropriate display units. All are to be displayed in miles.

8. *Insert* a **north arrow** on the layout page.

9. *Insert* your **name** and **date** on the layout page.

10. *Insert* a **neatline** as a border on the layout page.

11. *Export* the map as **S3SAL4_ XX** (where **XX** is your initials) in **JPEG** format to your **student folder**.

12. *Print* the **layout page**.

13. *Save* the project.

14. Allow your instructor to see your work before you exit ArcMap.

15. *Select* **Exit** from the **File** menu or *click* the **Close** button in the upper right corner of the ArcMap window to exit.

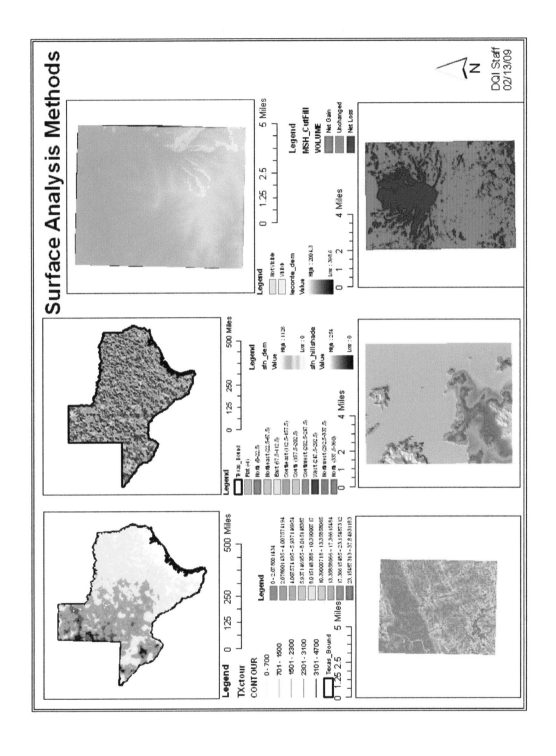

Surface Analysis Methods

Lesson 4: Enrichment Activity

In the earlier lesson, you gained experience creating elevation contours, determining the slope of the landscape, identifying the aspect of the landscape, creating a hillshade that shows terrain relief, evaluating visibility using viewshed analysis and estimating change in areas and volumes using cut/fill analysis from surface grid data. In this **Enrichment Activity**, you will apply many of these skills to your local digital elevation model (DEM).

Creating Contours

1.   *Launch* **ArcMap** with a **new blank map**.

2.   If ArcMap is still in **Layout View**, *click* to change to **Data View** .

3.   *Save* the project as **S3SAL4_Enrich_XX.mxd** (where XX is your initials) in your **student folder**

4.   *Add* the **YYYY_XXX_grd** grid (where **YYYY** is your county abbreviation and **XXX** is your school abbreviation) located in the **C:\STARS\ExtToolsinSA\local** folder. This is the DEM of your local USGS quadrangle.

   *Note: If your coordinate system is "unknown", you will need to build it by **right clicking** on the DEM layer, **selecting** **Predefined/Projected Coordinate Systems** and **selecting** the projection parameters contained in the UTM street clip shapefile name located in the **C:\STARS\ExtToolsinSA\local** folder. For example, for shapefile XXXX_YYY_str_utm27_z16n.shp, the projection parameters used to project the DEM are **UTM, NAD27, Zone 16N**.*

5.   *Select* **Surface Analysis ▶ Contour...** from the Spatial Analyst ▼ drop-down menu.

If the **Input height range** is close, a low **Contour interval** value will be defaulted; conversely, if the range is high, a higher **Contour interval** value will default.

---

6.   ***Enter*** a **Z factor** of **0.3048** if your DEM Z values are in feet (check a local map or internet reference to confirm the elevation range of your area and compare them to the DEM elevation values); or ***accept*** the default **1** if the Z value is already in meters.

7.   ***Accept*** the other **default values** for the analysis variables.

8.   ***Save*** the **output features** as **YourQuadName_Con.shp**.

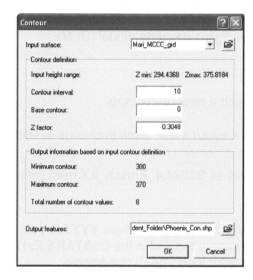

9.   ***Click*** [OK] to create the contours.

10.   ***Change*** the **symbology** so the **XXXX_YYY_grd elevation grid** data is displayed with the **Elevation #1 color ramp** .

11.   ***Double click*** the **YourQuadName_con** contours layer to open **Layer Properties**.

12. In **Symbology**, *specify* to show the contours using the **Graduated Colors** method to view the variation in elevation values.

13. *Select* the **CONTOUR** field as the **Value field**.

14. *Select* the **Green Bright color ramp** .

15. *Click* OK.

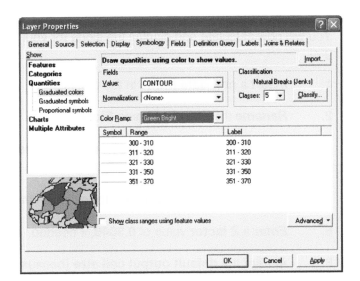

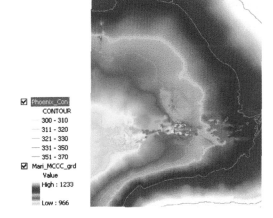

16. *Rename* the data frame **Contours**.

17. *Save* 🖫 the project.

<u>Calculating Slope</u>
You will now calculate and display the slope for your local quadrangle using the local elevation grid.

1.　　**Insert** a new **data frame** in the project.

2.　　**Rename** the new data frame **Slope**.

3.　　**Drag** the **YYYY_XXX_grd** grid into the **Slope** data frame from the **Contours** data frame to copy it.

4.　　**Select** Surface Analysis ▶ **Slope...** from the Spatial Analyst ▼ drop-down menu.

5.　　**Enter** a **Z factor** value of **0.3048**, if needed.

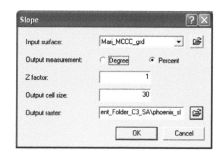

6.　　**Accept** the **default output cell size** (because this is the same cell size as the DEM grid)

7.　　**Save** the **output raster** as **YourQuadName_sl** in your **student folder**.

8.　　**Click** [OK] to create the slope raster.

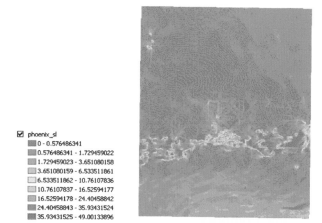

☑ phoenix_sl
　■ 0 - 0.576486341
　■ 0.576486341 - 1.729459022
　■ 1.729459023 - 3.651080158
　■ 3.651080159 - 6.533511861
　□ 6.533511862 - 10.76107836
　□ 10.76107837 - 16.52594177
　■ 16.52594178 - 24.40458842
　■ 24.40458843 - 35.93431524
　■ 35.93431525 - 49.00133896

9.　　**Save** 💾 the project.

Computing Aspect

1. *Insert* a new **data frame** in the project.

2. *Rename* it **Aspect**.

3. *Drag* the **YYYY_XXX_grd** grid into the **Aspect** data frame from the **Slope** data frame.

4. *Select* **Surface Analysis ▶ Aspect** from the
    Spatial Analyst ▼ drop-down menu.. Use the default cell
    size (which is the same cell size of the DEM grid)

5. *Save* the **output raster** as **YourQuadName_asp** in
    your **student folder**.

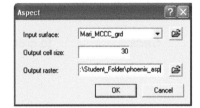

6. *Click* OK to create the aspect raster.

☑ phoenix_asp
  ☐ Flat (-1)
  ☐ North (0-22.5)
  ☐ Northeast (22.5-67.5)
  ☐ East (67.5-112.5)
  ☐ Southeast (112.5-157.5)
  ☐ South (157.5-202.5)
  ☐ Southwest (202.5-247.5)
  ☐ West(247.5-292.5)
  ☐ Northwest (292.5-337.5)
  ☐ North (337.5-360)

7. *Save* 💾 the project.

Displaying Hillshade

1. *Insert* a new **data frame** in the project.

2. *Rename* it **Hillshade**.

3. *Drag* the **YYYY_XXX_grd** grid located in the **Aspect** data frame.

---

4. *Select* **Surface Analysis ▶ Hillshade** from the `Spatial Analyst ▼` drop-down menu.

5. *Accept* the **default** values for **azimuth**, **altitude** and **output cell size**.

6. *Assign* a **Z factor** of *0.3048*, if needed.

7. *Save* the **output raster** as **YourQuadName_hs** in your **student folder**.

8. *Click* **OK** to create the hillshade layer.

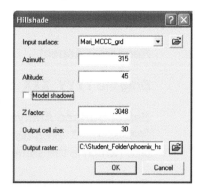

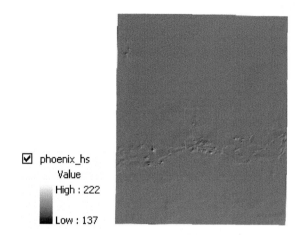

9. *Drag* the **DEM** data layer above the **hillshade** layer in the Table of Contents.

10. If the **Effects toolbar** is not currently visible in the ArcMap window, *select* **Toolbars ▶ Effects** from the **View** menu.

11. Make sure the **layer** on the **Effects toolbar** is set as the **XXXX_YYY_grd** DEM layer.

12. *Click* the **Adjust Transparency** button and *slide* the setting to **60%**.

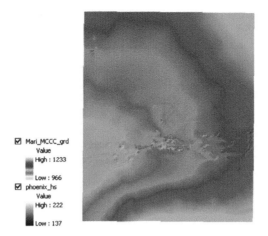

Mari_MCCC_grd
Value
High : 1233
Low : 966
phoenix_hs
Value
High : 222
Low : 137

13.   ***Save*** 🖫 the project.

Analyzing Viewshed

You will evaluate the viewshed from either a summit located in your local quadrangle or your school (depending on what data is available for your community and what geographic features exist in your community.)

1.   ***Insert*** a new **data frame** in the project.

2.   ***Rename*** it **Viewshed**.

3.   ***Drag*** the **XXXX_YYY_grd** grid from the **Aspect** data frame into the **Viewshed** data frame.

4.   ***Add*** ✚ the **XXXX_school.shp** data layer from the **C:\STARS\ExtToolsinSA\local** folder.

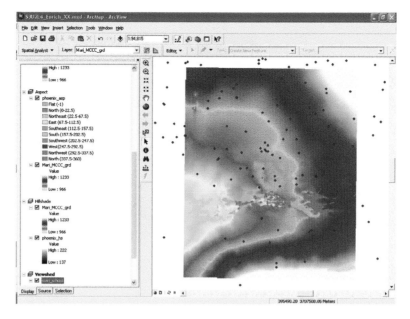

5.   Use the **Select by Attributes** method to *select* your school point.

6.   When you have made the selection, *close* the **Select by Attributes** dialog box.

7.   *Export* this selected feature as a new shapefile.

8.   *Save* the data file as **YourSchoolName.shp** in your **student folder**.

9.   *Click*   OK   to export the selected feature as a new shapefile.

10.　When prompted, **add** this new data file to the map.

11.　**Remove** the original **XXXX_school** school data layer.

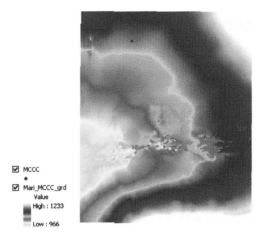

12.　**Select** Surface Analysis ▶ **Viewshed** from the Spatial Analyst ▼ drop-down menu.

13.　**Use** the new shapefile as the **Observer point**.

14.　**Input Z factor** of **0.3048**, if needed,

15.　**Use** an **output cell size** of **30**.

16.　**Save** the **output raster** as **YourQuadName_vs** in your student folder.

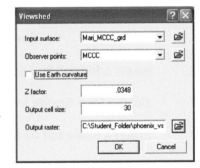

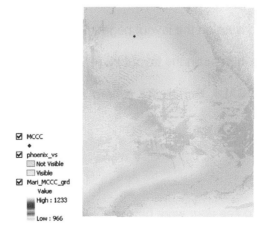

17.　In the Table of Contents, **uncheck** the **XXXX_YYY_grd** DEM layer in the **Slope** and **Aspect** data frames.

---

18.   **Save** 🖫 the project.

Creating the Map Layout:
You will now create a map layout to display the analysis results from this enrichment activity.

1.   **Switch** to **Layout View** by clicking on the **Layout View** ▣ ◻ button at the bottom of the Data View.

2.   **Change** the page orientation the **Page Orientation** to **Landscape**.

3.   **Position** the data frames by dragging them to new locations on the layout using your mouse.

4.   **Resize** the data frames on the map layout page by clicking on the data frame boxes and using the sizing handles to click and drag.

5.   **Insert** the following **title** on the layout page in **24-point bold** type:

Surface Analysis – Enrichment Exercise

6.   **Insert** a **legend** for each data frame on the map layout page.

7.   **Set** the **Map Scales** for each of the data frames on the layout page as the **same value**.

8.   **Insert** a **scale bar** on the layout page using appropriate display units.

9.   **Insert** a **north arrow** on the layout page.

10.   **Insert** your **name** and **date** on the layout page.

11.   **Insert** a **neatline** as a border on the layout page.

12.   **Export** the map as **S3SAL4_Enrich_XX** (where **XX** is your initials) in **JPEG** format to your **student folder**.

13.   **Print** the **layout page**.

14.   **Save** 🖫 the project.

15.   Allow your instructor to see your work before you exit ArcMap.

16.   **Select Exit** from the **File** menu or **click** the **Close** ☒ button in the upper right corner of the ArcMap window to exit.

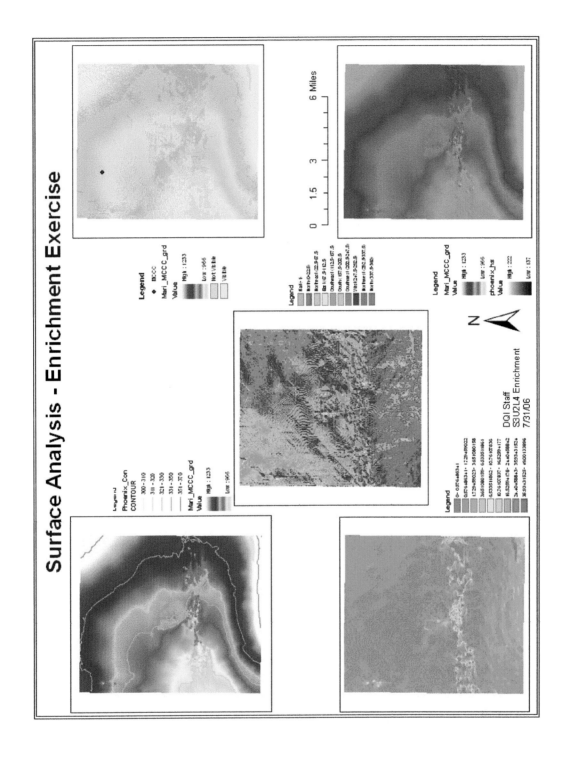

Lesson 4: Analyzing Surfaces Lesson Review

Key Terms
Use the lesson or index provided in the back of the book to define each of the following terms.

1. Contours
2. Contour Interval
3. Base Contour
4. Slope
5. Aspect
6. Hillshade
7. Viewshed
8. Cut/Fill Analysis

Global Concepts
Use the information from the lesson to answer the following questions. You may need to answer these on the back of this page or on your own paper.

9. Could you create contours and slope calculations for something besides land? If so, give an example.

10. What is the advantage of showing a hillshade analysis to somebody unfamiliar with GIS rather than some of the others in today's exercise?

11. Cut/fill analysis can be very useful. Give an example besides the exercise of how it could be used.

Let's Talk About It...
Answer the following question and share the responses with your instructor and classmates.

12. What fields of work might require regular usage of surface analysis and why?

Lesson 5:
Creating Grid Statistics

In analyzing statistics for raster data, you can perform various different calculations depending on the kind of information you need. Spatial Analyst can be used to calculate cell statistics, neighborhood statistics or zonal statistics.

Cell statistics are calculations made about each location, or cell, of a raster grid, among multiple rasters of the same geographic area. This enables you to see how data values (such as land use, elevation, etc.) of individual cells, or locations, change over a period of time.

The majority calculation determines the value that occurs most often on a cell-by-cell basis for the input grids. An example of a majority calculation would be if somebody was looking at average household income levels in an area, they could find out the average household income that most cells in an area show, which would give the average household income that a majority of the cells had.

The maximum calculation determines the maximum value on a cell-by-cell basis. Again using the example of average household income, this calculation would take the largest household income in the cells in the area, and show the largest household income in each cell. The mean calculation computes the mean of the values on a cell-by-cell basis. So for the average household income, it would average the averages together and show that value.

The calculation of the median computes the median of the values on a cell-by-cell basis. Median is the value that occurs in the middle of the list of values. So in the case of the average household incomes across cells, all the household income values would be lined up by value and whatever was in the middle would be the median value displayed(ie if there were 11 values, the sixth value, whatever it was, would be the median). The calculation of the minimum determines the minimum value on a cell-by-cell basis. For the household income example, whatever household income was the lowest would be the displayed value. The calculation of the minority determines the value that occurs least often on a cell-by-cell basis. Utilizing the average household income numbers again, the minority value would be the household income which occurred least often.

The range is calculated to determine the range of values on a cell-by-cell basis. With the household income example, the range value shows what the highest and lowest household income within the area, so you would see how much it varied.

Standard deviation is used to compute the standard deviation of the values on a cell-by-cell basis. Standard deviation is a value found in statistics, and is a meta-analysis – rather than

telling you about the household incomes themselves, it talks about the variations in the household incomes. The standard deviation values you would end up with tell you how many standard deviations there are between different cells.

The sum calculation computes the sum of the values on a cell-by-cell basis. With the example of the household average income, it adds up all the household incomes in the cell and presents that as the cell value.

The Variety is calculated to determine the number of unique values on a cell-by-cell basis. This is another meta-analysis like standard deviation. It will tell you how many different average household incomes there were, so instead of knowing what the incomes were, it tells you whether there was a lot of different average incomes in the area or if the household incomes were pretty much the same. The higher the variety, the more variable the incomes were.

Neighborhood statistics are calculations made about groups of cells to see how data values vary in a particular cell neighborhood. Neighborhood statistics uses a shape to identify the extent of the neighborhood. This shape can be a rectangle, or any polygon that contains four 90 degree angles, a circle, an annulus, or a wedge. Just like the real world have different shaped areas that groups of people live in, such as the shape of a street or an area of several blocks in a city with similar backgrounds, a neighborhood of cells can be defined by different shapes. Once a neighborhood is defined, you have that much change showed. For example if your neighborhood had 9 cells in it, you would have nine different degrees of change – the real world equivalent is that if a neighborhood had 9 different houses in it, each house would be unique in some way.

Zonal statistics helps to identify if relationships exist between landscape features. The analysis in the exercise identifies if a relationship exists between land use variety and slope in the area. There is a limited number of land use types, and the question with zonal statistics is whether there is a greater variety in the land use depending on how steep the slope was. In the analysis you will see in the following exercise, the more the slope varied, the more land use types were in that zone.

Lesson 5: Creating Grid Statistics

In analyzing statistics for raster data, you can perform various different calculations depending on the kind of information needed in your analysis. **Spatial Analyst** can be used to calculate **cell statistics**, **neighborhood statistics** or **zonal statistics**.

Cell Statistics – Analyzing Precipitation Data

Cell statistics are calculations made about each location, or cell, of a raster grid, among multiple rasters of the same geographic area. This enables you to see how data values (such as land use, elevation, etc.) of individual cells, or locations, change over a period of time.

To use **Cell Statistics** to determine the total annual precipitation in North Carolina,

1.   *Launch* ArcMap with a **new blank map**.

2.   *Save* 💾 the map document as **S3SAL5_XX.mxd** (where **XX** is your initials).

3.   *Add* ✛ the **nc_prec_sum** and **nc_prec_win** raster grids located the **C:\STARS\ExtToolsinSA**folder. These raster grids display seasonal precipitation totals for summer and winter in North Carolina.

Note that the precipitation values for summer range from 0 to 19 inches, and winter from 0 to 6 inches.

4. **Zoom** 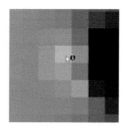 into a specific area of the raster grid to see the individual cells.

5. **Use** the **Identify Tool** 🛈 view the value of one cell as you **toggle** each raster grid layer on and off.

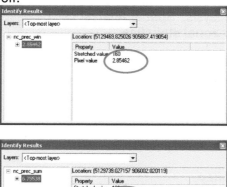

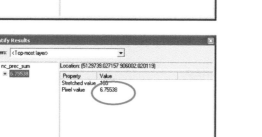

6. **Note** the sum of these two values.

7. To create a new layer of the combined sum of each cell for summer and winter precipitation totals, **select** **Cell Statistics...** from the Spatial Analyst ▼ drop-down menu.

8. **Highlight** the **nc_prec_sum** and **nc_prec_win** raster grid layers in the **Layers** list.

9. **Click** [Add ->] to specify them as **Input rasters**.

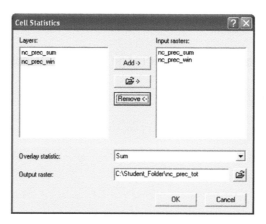

10. From the **Overlay statistic** drop-down list, **select** **Sum**.

Other options for overlay statistics for Cell Statistics calculations include:

Type of cell statistic	Description
Majority	Determines the value that occurs most often on a cell-by-cell basis for the input grids
Maximum	Determines the maximum value on a cell-by-cell basis
Mean	Computes the mean of the values on a cell-by-cell basis
Median	Computes the median of the values on a cell-by-cell basis
Minimum	Determines the minimum value on a cell-by-cell basis
Minority	Determines the value that occurs least often on a cell-by-cell basis
Range	Determines the range of values on a cell-by-cell basis
Standard deviation	Computes the standard deviation of the values on a cell-by-cell basis
Sum	Computes the sum of the values on a cell-by-cell basis
Variety	Determines the number of unique values on a cell-by-cell basis

(Source: ArcGIS Desktop Help)

11.   *Save* the **Output raster** as **nc_prec_tot** in your **student folder**.

12.   *Click* **OK**. The new raster grid is added to the map.

Each cell in this raster grid is the sum of each precipitation value in the original two raster grids.

13.   *Check* the value of the same cell that you viewed in Step 5 to confirm that it is the sum of both values.

14.   To change the classification intervals for this new grid to better display variation in this data, *double click* it to open **Layer Properties**.

15.   *Click* the **Symbology** tab.

16.   *Confirm* that the display method is set as **Classified**.

17.   *Change* the **Color Ramp** for the classification to ▬▬▬▬▬ (**Precipitation**).

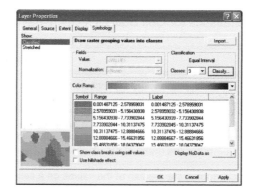

18. ***Click*** [OK] to **apply** the changes and ***close*** **Layer Properties**.

This new map displays the total precipitation for North Carolina, calculated using the cell statistics tool, to sum the seasonal totals for summer and winter.

19. ***Uncheck*** the **nj_prec_sum** and **nc_prec_win** layers in the Table of Contents.

20. ***Rename*** the data frame **Layers** to **Cell Statistics** and ***collapse*** it. ⊞ ≋ Cell Statistics

21. ***Save*** the project.

Neighborhood Statistics

Neighborhood statistics are calculations made about **groups of cells** (a neighborhood) in a raster grid. This enables you to see how data values vary in a particular cell neighborhood.

Neighborhood statistics uses a shape to identify the extent of the neighborhood. This shape can be a rectangle, or any polygon that contains four 90° angles, a circle, an annulus or a wedge.

(Source: ArcGIS Desktop Help)

Computations are run on all of the cells within each neighborhood, to determine one of the following statistics:

Type of neighborhood statistic	Description
Majority	Determines the value that occurs most often in the neighborhood
Maximum	Determines the maximum value in the neighborhood
Mean	Computes the mean of the values in the neighborhood
Median	Computes the median of the values in the neighborhood
Minimum	Determines the minimum value in the neighborhood
Minority	Determines the value that occurs least often in the neighborhood
Range	Determines the range of values in the neighborhood
Standard deviation	Computes the standard deviation of the values in the neighborhood
Sum	Computes the sum of the values in the neighborhood
Variety	Determines the number of unique values within the neighborhood

(Source: ArcGIS Desktop Help)

For example, if a **rectangle** of 3 cells by 3 cells (in this case, a square!) is chosen as the neighborhood, and **Sum** is chosen as the desired statistic, then the value of the central cell in the neighborhood is the sum total of all of the values of the cells surrounding it.

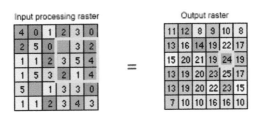

(Source: ArcGIS Desktop Help)

In this activity, you will analyze land use data for Cook County, Illinois which encompasses the Chicago Metropolitan Area to determine the diversity, or **variety**, of land uses in the county.

1. *Insert* a new **data frame.**

2. *Rename* it **Neighborhood Statistics**.

3. *Add* ⬇ the **cook_lulc** raster grid located the **C:\STARS\ExtToolsinSA** folder.

4. *Double click* the layer to open **Layer Properties**.

5. *Select* to display the data using the **Unique values** method.

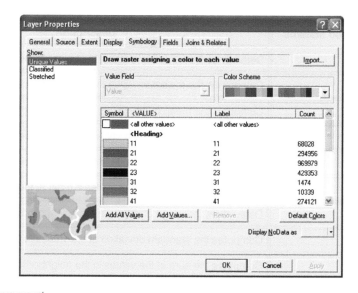

6. **Click** [OK] to apply the change and close **Layer Properties**.

The colors of your map may vary from these colors, as they are randomly generated when added to the map. The map shows a list of classification codes assigned by the USGS to delineated land use and land cover:

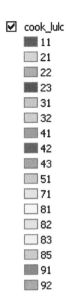

☑ cook_lulc
◼ 11
☐ 21
▨ 22
◼ 23
☐ 31
▨ 32
◼ 41
◼ 42
▨ 43
☐ 51
☐ 71
☐ 81
☐ 82
☐ 83
▨ 85
◼ 91
▨ 92

To determine how diverse the county is with respect the land use and land cover, we will run a neighborhood statistic function to report how many unique land uses are contained within a defined cell neighborhood.

7. ***Select* Neighborhood Statistics…** from the Spatial Analyst ▼ drop-down menu.

8. ***Confirm*** the **cook_lulc** raster grid as the **Input data**.

9. ***Select* Variety** as the **Statistic type**

10. ***Select* Rectangle** as the **Neighborhood**.

11. ***Accept*** the **default height**, **width**, **units** and **Output cell size**.

12. ***Save*** the **Output raster** as **Cook_LUV** in your **student folder**.

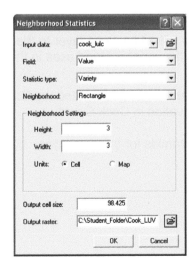

13. **Click** ☐ OK ☐. When processing is complete, the variety raster grid appears in the map display.

The parameters set to run the neighborhood statistic on each cell in the Cook County land use/land cover raster grid determines how many unique land use/land cover types are contained within the 3 cell by 3 cell neighborhood surrounding it. Each cell is then assigned that value up to a maximum possibility of nine (3x3). The higher the resulting variety value, the more diverse the defined neighborhood!

In this case, the lighter the shade of purple, the more diverse the land uses. For Cook County, the area along the right perimeter of the county shows the least diversity, as this is the major metropolitan area of Chicago proper, and contains a high concentration of urban commercial and residential land uses.

14. **Save** this project.

Zonal Statistics
Running statistics on the zone data helps to identify if relationships exist between landscape features. In this example, you will add a slope layer to the analysis to see if this landscape feature has any correlation to the variety of land uses in the study area.

1. **Add** ⬇ the **cook30dem** raster grid located the **C:\STARS\ExtToolsinSA**folder. This raster grid displays elevation in Cook County.

2. If prompted to build pyramids for the data layer, **click** ☐ Yes ☐.

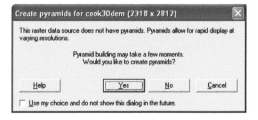

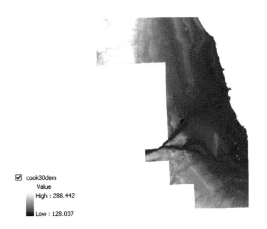

☑ cook30dem
Value
High : 288.442

Low : 128.037

3. **Create** a **slope** layer from this elevation raster grid layer

4. **Save** it in your **student folder** as **cook_slope**.

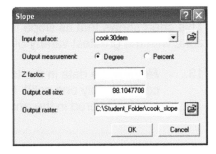

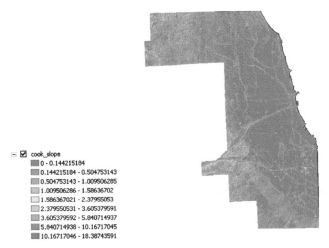

⊟ ☑ cook_slope
■ 0 - 0.144215184
■ 0.144215184 - 0.504753143
■ 0.504753143 - 1.009506285
□ 1.009506286 - 1.58636702
□ 1.586367021 - 2.37955053
□ 2.379550531 - 3.605379591
■ 3.605379592 - 5.840714937
■ 5.840714938 - 10.16717045
■ 10.16717046 - 18.38743591

5. **Select** Zonal Statistics... from the Spatial Analyst ▼ drop-down menu.

6. **Select cook_luv** as the **Zone** dataset;

7. **Click Value** as the **Zone** field;

8. **Use cook_slope** as the **Value** raster;

9. **Check on** to **Ignore NoData in calculations** and **Chart statistic**;

---

10.   *Select* **Mean** as **Chart Statistic**;

11.   *Save* the **Output table** as **cook_zone_slope.dbf** in your **student folder**.

12.   *Click* 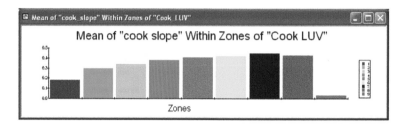. The zonal statistics chart and table appear.

This chart shows that as slope increases, the variety of land uses also increases, except for the zone with the greatest variety of land uses.

13.   *Review* the data in the table to see that this one zone with the greatest variety only contains only one cell!! Because this one cell is so insignificant to the overall dataset, it can be ignored in the analysis.

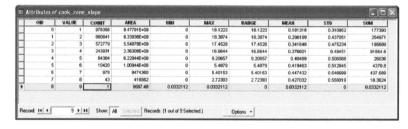

By generating statistics for each of these datasets, more meaningful and effective information has been extracted to enhance analysis of the study areas.

14.   *Close* the **graph window** and the **table**.

15.   *Uncheck* the **cook_luv** and **cook_lulc** layers in the Table of Contents.

16.   *Save* the project.

Creating the Map Layout:
1.   *Switch* to **Layout View**.

2.   *Change* the **Page Orientation** to **Landscape**.

3.   *Insert* the following **title** on the map layout in **24-point bold** type:

Cell, Neighborhood & Zonal Statistics

4.   *Insert* a **legend** for each data frame on the map layout page.

5. *Insert* a **scale bar** on the layout page for each data frame using display units that are appropriate for the map layout.

6. *Insert* a **north arrow** on the layout page.

7. *Place* your **name** and the **date** on the layout page.

You will now place the graph that you created on the map layout page.

8. *Select* **Graphs ▸ Mean of "cook_slope" Within Zones of "cook_luv"** from the **Tools** menu. The graph will open.

9. *Right click* the **graph title bar** 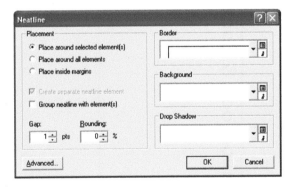 and *select* **Show on Layout**. The graph will appear on the layout page.

10. *Close* the **graph** window.

11. *Drag* the **graph** to an appropriate location on the layout page.

    You will now add a neatline around the graph.

12. *Click* to select the **graph** on the layout page.

13. *Select* **Neatline...** from the **Insert** menu.

14. Make sure the
    ⦿ Place around selected element(s) option is
    selected.

15. *Select* a **single**, **1-point** border style for the neatline.

16. *Click* OK .

17. *Insert* a **neatline** as a border on the map layout page.

18. *Export* the map as **S3SAL5_XX** (where **XX** is your initials) in **JPEG** format to your **student folder**.

19. *Print* the layout.

20. *Save* 💾 the project.

21. Allow your instructor to see your work before you exit ArcMap.

22. *Exit* ArcMap.

Cell, Neighborhood & Zonal Statistics

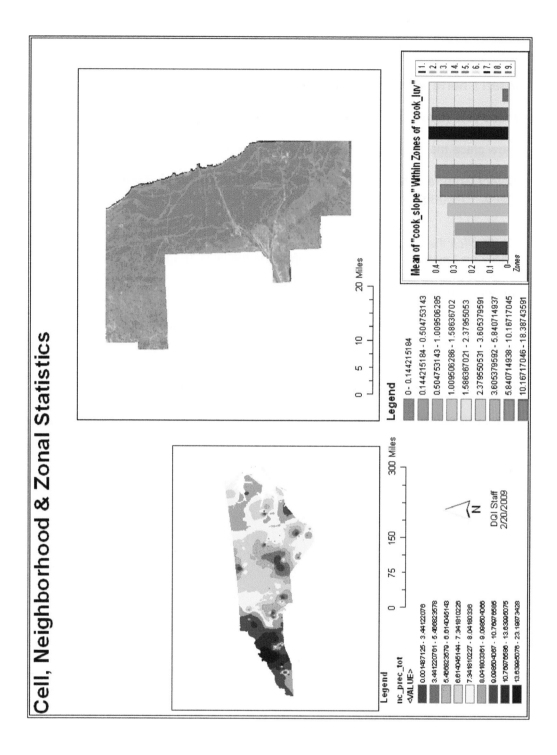

Lesson 5: Enrichment Activity

In the previous exercise, you learned how to calculate cell statistics, neighborhood statistics and zonal statistics of raster grid data using **Spatial Analyst**. You will now apply these skills to your local data set. Because your local data set does not contain two raster grid layers that could be used effectively to compute cell statistics, this exercise will be limited to neighborhood and zonal statistics.

Neighborhood Statistics

1. *Launch* **ArcMap** with a **new blank map**.

2. If ArcMap is still in **Layout View**, *click* to change to **Data View** .

3. *Save* the project as **S3SAL5_Enrich_XX.mxd** (where XX is your initials) in your **student folder**

4. *Add* the **YYYY_XXX_grd** data layer located in the **C:\STARS\ExtToolsInSA\local** (where **YYYY** is your county abbreviation and **XXX** is your school abbreviation) folder.

5. *Add* the **YourQuadName_lulc** grid data layer located in your **student folder** that you created in **Lesson 1 Enrichment Activity**.

6. *Add* the **LULC.txt** table located in the **C:\STARS\ExtToolsInSA\local** folder.

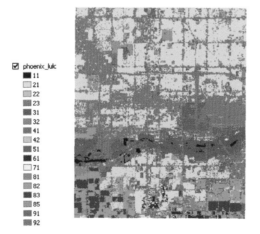

7. *Open* the *attribute* table for the **YourQuadName_lulc** layer.

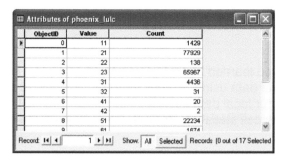

Notice that the table contains the values for the land use for your county (**Value** field) but does not contain the descriptions of these codes.

8.   ***Close*** the attribute table.

9.   ***Open*** the ***LULC.txt*** table.

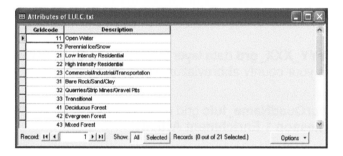

This table contains the descriptions of the land use codes. You will join this table to the land use grid layer.

10.   ***Close*** the table.

11.   *Right click* the **YourQuadName_lulc** data layer and *select* **Joins and Relates ▶ Join**.

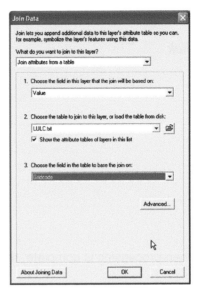

12.   In the **Join Data** dialog box, *select* the **Value** field as the field on which to base the join.

13.   *Select* the **LULC.txt** table as the table to join

14.   *Select* **Gridcode** as the field in the table to use for the join.

15.   *Click*   **OK**   .

16.   To check the result of the join, *open* the **attribute table** for the **YourQuadName_lulc** layer again. The data layer table now contains the additional data from the text table.

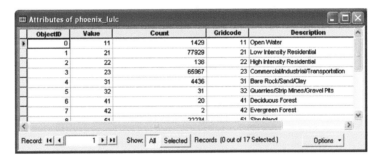

17.   *Close* the table.

You will now use **Neighborhood Statistics** to determine the variety of land uses in your community.

18.   *Select* **Neighborhood Statistics...** from the Spatial Analyst ▼ drop-down menu.

19.   *Confirm* the land use grid that you created as the **Input data** box.

20.   *Confirm* **Value** as the **field**.

21.   *Specify* **Variety** as the **Statistic type**.

22.   *Confirm* **Rectangle** as the **Neighborhood**.

23.   *Accept* the **default height**, **width**, **units** and **Output cell size**.

24.   *Save* the **Output raster** as **YourQuadName__luv** in your **student folder**.

25. **Click** OK .

When processing is complete, the variety grid appears in the map display.

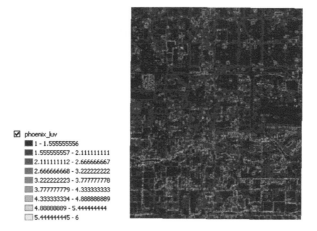

As the legend in the table of contents displays, the lighter the shade of purple, the more diverse the land uses.

26. **Save** 💾 the project.

Zonal Statistics

You can now calculate a slope grid to find out if a relationship exists between the slope of the landscape and the variety of land uses in your community.

1. **Add** ✛ the **YourQuadName_Slope** grid data layer from your **student folder** that you created in **Lesson 1 Enrichment**.

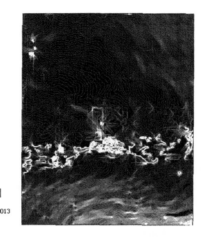

phoenix_slope
Value
High : 49.0013

Low : 0

2.   ***Select* Zonal Statistics...** from the Spatial Analyst ▼ drop-down menu.

3.   ***Set*** the **Zone dataset** as the **YourGridName_luv** variety grid that you created earlier.

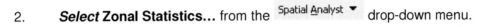

4.   ***Specify* Value** as the **Zone field**

5.   ***Select* YourQuadName_Slope** as the **value raster**.

6.   ***Select* Mean** as the **chart statistic**.

7.   ***Save*** the **output table** as **zstat_YourQuadName.dbf** in your **student folder**.

8.   ***Click*** OK . The zonal statistics chart and table appear.

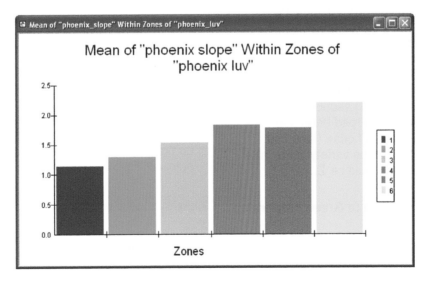

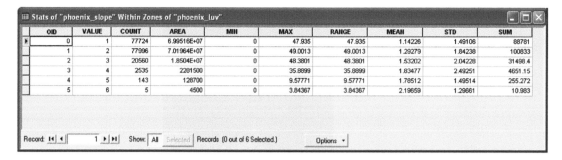

Is there any correlation between slope and land use variety for your quad? Are there any zones that contain only a few cells that may skew the distribution, as was the case in the earlier lesson?

9.   ***Close*** the table and the chart.

10.   In the Table of Contents, ***uncheck*** the **YourQuadName_slope**, **YourQuadName_lulc** and **YYYY_XXX_grd** data layers.

11.   To better display the **YourQuadName_luv** variety layer, ***double click*** the layer to open **Layer Properties**.

12.   ***Click*** the **Symbology** tab.

13.   ***Specify*** to **show** the data using the **Unique Values** method.

This method will display the variety of land uses in integer values instead of decimal values that are more difficult to understand when evaluating variety. If you were evaluating a 30-meter by 30-meter section of land, you would determine the number of land uses incorporated into that section of land in whole numbers. For example, one (1) section of land could contain a road (land use code 23) and a house (land use code 21). So the variety value in that section would be 2.

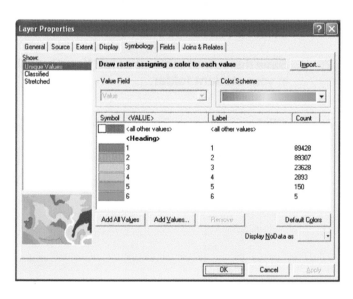

14.   ***Select*** the **Red to Green color ramp** .

15.   ***Click***   OK   to apply the change and close **Layer Properties**.

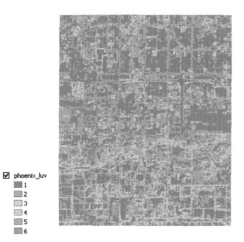

16.   ***Save*** 💾 the project.

Creating the Map Layout:

You will now create a map layout to display the analysis results from this enrichment activity.

1.   ***Switch*** to **Layout View**.

2.   ***Change*** the **Page Orientation** to **Landscape**.

3.   ***Insert*** the following **title** on the map layout in **24-point bold** type:

         Neighborhood & Zonal Statistics – Enrichment Exercise.

4.   ***Insert*** a **legend** on the map layout page that includes the **YourQuadName_luv** variety raster only.

        The legend will appear on the map layout page as follows:

Legend
- ▣ 1
- ▣ 2
- □ 3
- ▢ 4
- ▣ 5
- ▣ 6

        This legend does not explain enough about the data contained in the map.

5.   **Double click** the **legend** on the layout page to open **Legend Properties**.

6.   **Select** the **YourQuadName_luv** data layer in the **Legend Items** list.

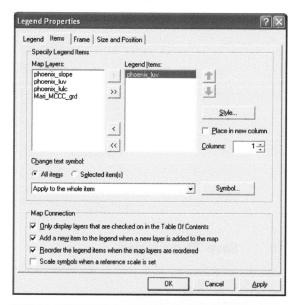

7.   **Click** the [Style...] button to open the **Legend Item Selector**.

8.   **Select** the

Legend

Layer Name
Heading
■■■ Label

Horizontal with Layer Name, Heading and Label

style so that the layer name will be included at the top of the symbology in the legend.

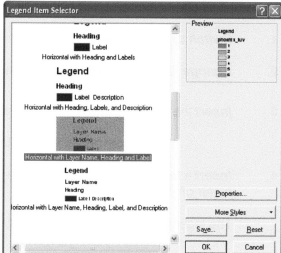

9.   **Click** [OK] to close the **Legend Item Selector** and return to **Legend Properties**.

10.   **Click** [OK] to apply the change and close **Legend Properties**.

11.   When the legend is added to the map layout, **resize** it and **move** it to an appropriate place on the layout page.

12.   **Insert** a **scale bar** on the layout page using display units that are appropriate for the map layout.

13.   **Insert** a **north arrow** on the layout page.

14.   **Place** your **name** and the **date** on the layout page.

You will now place the graph that you created on the map layout page.

15. **Select Graphs ▶ Mean of "YourQuadName slope" Within Zones of "YourQuadName luv"** from the **Tools** menu. The graph will open.

16. **Right click** the **graph title bar** 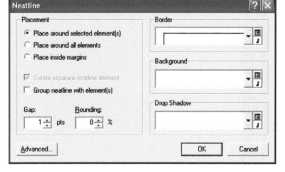 and **select Show on Layout**. The graph will appear on the layout page.

17. **Close** the **graph** window.

18. **Drag** the **graph** to an appropriate location on the layout page.

    You will now add a neatline around the graph.

19. **Click** to select the **graph** on the layout page.

20. **Select Neatline...** from the **Insert** menu.

21. Make sure the  option is selected.

22. **Select** a **single**, **1-point** border style for the neatline.

23. **Click** OK .

24. **Insert** a **neatline** as a border on the map layout page.

25. **Export** the map as **S3SAL5_Enrich_XX** (where **XX** is your initials) in **JPEG** format to your **student folder**.

26. **Print** the layout.

27. **Save** 🖫 the project.

28. Allow your instructor to see your work before you exit ArcMap.

29. **Exit** ArcMap.

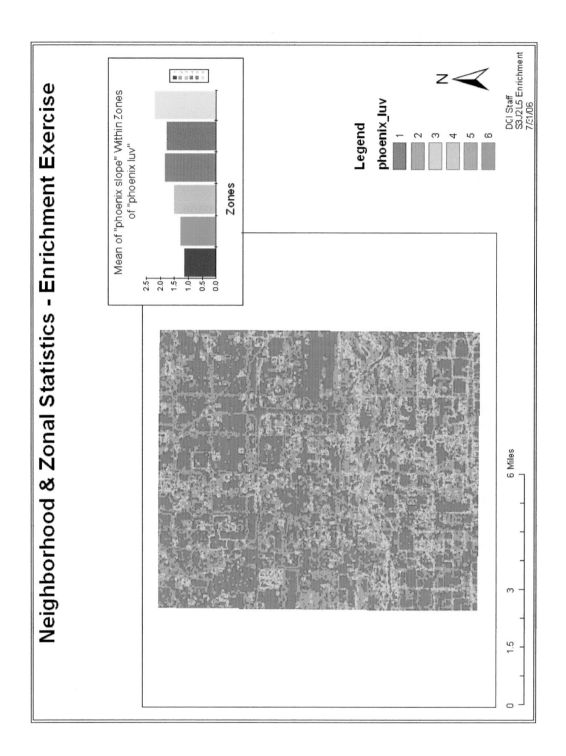

Neighborhood & Zonal Statistics - Enrichment Exercise

Lesson 5: Creating Grid Statistics Lesson Review

Key Terms
Use the lesson or index provided in the back of the book to define each of the following terms.

1. Cell Statistics
2. Neighborhood Statistics
3. Zonal Statistics

Global Concepts
Use the information from the lesson to answer the following questions. You may need to answer these on the back of this page or on your own paper.

4. Besides precipitation, give another example of another dataset that you might want to look at cell statistics for. Also note down what type of cell statistic you would calculate and why.

5. If you did a neighborhood statistic calculation looking for the minority, what kind of data might you be working with, and why?

6. Give an example of information you might want to study the zonal statistics of, and explain why.

Let's Talk About It...
Answer the following question and share the responses with your instructor and classmates.

7. What fields of work might require regular usage of spatial statistics and why?

Lesson 6:
Applications - Surface Analysis Project

The previous exercises were designed for you to use sample data sets to see the tools available for you to use. This exercise is different in that your goal here is to see how these tools can be used for any project. This exercise will take you through an example problem that you will resolve using your local data set. For this exercise you will be looking at land use data and analyzing it. This project will allow you to see some of the many ways that GIS is used in careers. The analysis work you will be doing here is done regularly by developers, planners, and other professionals working with land management or economic development. For example the work done in here would be useful for a planner or somebody on a zoning board to aid them in determining how pieces of property should be zoned or utilized by people. Also, a developer or other person in the business of buying and utilizing land, might also want this information to determine whether or not a piece of property would be a cost-effective purchase for them, or if they already own the property, allow them to predict potential problems and be able to avoid or work around them before they become a problem or a financial loss.

When performing analysis using Spatial Analyst, it is important to utilize options available to enhance and, when possible, simplify analysis. In Spatial Analyst Options, you can specify a particular data layer's extent to use as the extent of the analysis function. By setting the Analysis Extent, you specify the four (4) x,y coordinates that will form the boundaries of the area you want to analyze. In Options, you can also specify the cell size of the output grid that will be formed from the analysis. The cell size is described differently depending on the coordinate system you are using, so if you are working in UTM for example, your cell size would be defined in meters. In the case of using your UTM coordinate system then, a cell size of 1 would be 1 meter by 1 meter resolution, while if you were using a state plane system that measured in feet, a cell size of 1 would be 1 foot by 1 foot, which is about 3 times as detailed if you are measuring the same area. So be sure you know what coordinate system you are using and how fine you want your cell size to be.

For some projects, you may need land use data for only a portion of an area that has uneven boundaries, such as within the boundaries of a town or city. In this exercise you will evaluate the land use data within your city boundary. This is something that a planner or other city employee might be doing – creating a focus on their area of interest and doing the analysis around that. This same concept can be used for conservationists on focusing either within a park or other preserved piece of land to plan the best use for what they already own, or looking within the county or state they live or work in to find areas they need to try to conserve.

For this study, we will analyze the relationship between waterbodies/wetlands, and elevation. Developments in these areas run the risk of flooding, unstable soil, and wildlife habitat destruction. By identifying those areas where these risks may occur, planners and developers can protect them from inappropriate and dangerous development. For a planner or a developer, they would want to avoid building in the wetland area, because the process would be more difficult than on flatter, higher ground. By the same token, conservation groups might be more concerned about wetland areas to protect and preserve them and their surrounding areas to protect endangered species or prevent erosion and damage to waterways further downstream. The primary issue between the groups can be that while the planners or developers might only need a small amount of space around a water body or wetland area, there might be as much damage done if they build close to the sites as if they built on the sites.

In this graphic, the green area represents areas where none of the criteria has been met. The purple represents areas where the criterion has been met. In this case, it includes all areas that are either water or wetlands, AND are at or below 20 feet in elevation.

In this case, all of the areas in purple meet the land sensitivity criteria and must be excluded from any consideration for development.

A mask can be created to overlay on top of a detailed map or site plan of the area.

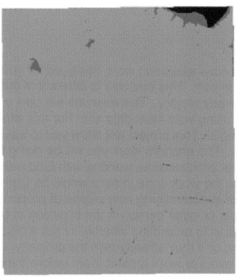

This map highlights the excluded areas, and helps planners and developers when analyzing potential sites for future development

Lesson 6: Applications – Spatial Analyst Project

You have already gained experience working with the **Spatial Analyst** extension. So far, you have used all of the various tools and techniques that are available with **Spatial Analyst.** In this exercise, you will use the **Raster Calculator** for two different purposes. First, you will use it as a tool to **clip** the land use data to the extent of your city boundary. Next, you will use the **Raster Calculator** to query land use and elevation data to determine **sensitive** areas in your community.

Analyzing City Land Use

You have used your local land use data numerous times in past lessons. In an earlier lesson, you clipped the land use data to the extent of the local quadrangle. For some projects, you may need land use data for only a portion of an area that has uneven boundaries, such as within the boundaries of a town or city. In this exercise you will evaluate the land use data within your city boundary.

1. *Launch* **ArcMap** with a **new blank map** and *save* the project as **S3SAL6_XX.mxd** (where **XX** is your initials) in your **student folder**.

2. *Add* the **YourQuadName_lulc** raster grid located in your **student folder** that you created in **Enrichment Lesson 1**.

3. *Add* **YYYY_trg##### plc00.shp** file (where **YYYY** is your county abbreviation) located in the **C:\STARS\ExtToolsinSA\local**folder.

4. *Click* the **Zoom to Full Extent** button .

5. *Click* the **Select Features** tool and *click* the polygon that represents your city in the map display. It will be highlighted in light blue.

6. *Export* your city polygon as a new shapefile.

7. *Save* it as **YourCityName.shp** in your **student folder**.

8. When prompted to add the new data layer to the map, *click* Yes .

9. *Remove* the original **YYYY_trg##### plc00.shp** data layer from the Table of Contents.

Since we are only interested in doing analysis on the land use data in your city, we need to clip the full land use raster grid down to only include the area inside your city boundary. In order to do this, your city shapefile needs to be converted to a raster grid layer.

10. *Select* **Convert ▶ Features to Raster** from the Spatial Analyst ▼ drop-down menu.

11.   *Set* the **Input features** to **your city shapefile**.

12.   *Select* **NAME** as the **Field**.

13.   *Accept* the **output cell size**.

14.   *Save* the **Output raster** as **YourCityName** to your **student folder**.

The new raster grid layer of your city is added to the map.

15.   *Remove* the city shapefile from the Table of Contents.

16.   To set the extent to only your city boundary, *select* **Options** from the Spatial Analyst ▼ drop-down menu.

17.   *Set* the **YourCityName** as the **Analysis mask**.

18.	From the **Extent** tab, *set* Same as Layer "**YourCityName**" as the **Analysis extent**.

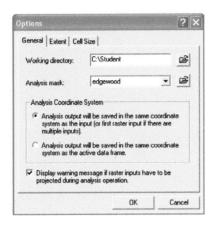 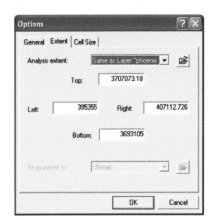

19.	***Click*** ⬚OK⬚ .

To clip the land use raster grid to the extent of your city boundary,

20.	***Select* Raster Calculator…** from the Spatial Analyst ▼ drop-down menu.

In the **Raster Calculator** query box,

21.	***Double click*** the **YourCitylulc** raster grid in the **Layers** list to add the layer name to the expression box.

By doing this, you are setting **Spatial Analyst** to display the same data that is already displayed in the LULC raster grid, except that by selecting the analysis mask and extent you specify to only display this data in the areas included in the mask, in this case, within your city boundary.

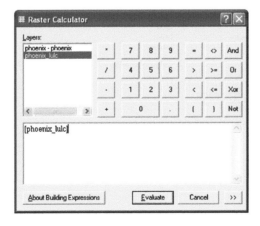

22.	***Click*** ⬚Evaluate⬚ . The **Calculation** raster grid is added to the map.

23.	***Remove*** the other layers in the Table of Contents.

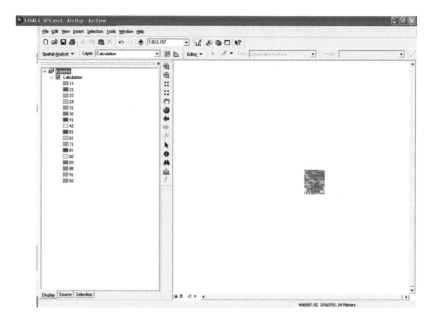

24.   *Open* **Layer Properties** for the new **Calculation** raster grid layer.

25.   *Click* the **Source** tab.

26.   *Scroll* down to notice that the **status** of this layer is **Temporary**.

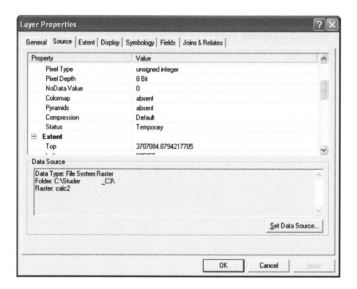

27.   *Close* **Layer Properties**.

28.   To make this layer permanent, *right click* the **Calculation** layer in the Table of Contents

29.   *Select* **Make Permanent...**

30.   In the **Make Calculation Permanent** dialog box, *navigate* to your **student folder**.

31.   *Name* this layer **YourCityName_LULC**.

At this point, the data is displayed in this new raster grid based on land use codes. In order to display the raster grid to show the land use descriptions you must add a table that contains this data and join that table to the grid.

32.   *Add* 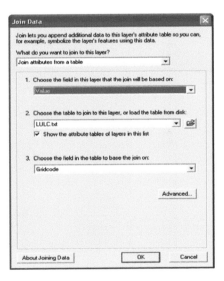 the **LULC.txt** table located in the **C:\STARS\ExtToolsinSA\local** folder.

33.   To join the **Calculation** raster grid to the text table, *right click* the **Calculation** raster grid layer in the **Table of Contents**

34.   *Select* **Joins and Relates ▶ Join...**

35.   *Select* the **Value** field as the field that the join will be based on for the **Calculation** layer.

36.   *Select* the **LULC.txt** layer as the table to join to the layer and **Gridcode** as the field in that table to use for the join.

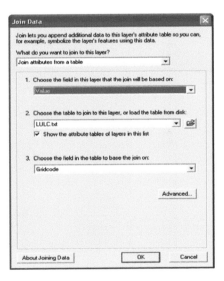

37.   *Click*   OK   .

38.   To map the city land use raster grid by the description of the type of land use, *open* **Layer Properties**

39.   *Click* the **Symbology** tab.

40.   *Select* **Unique Values**.

41.   *Select* **Description** as the **Value Field**.

42.   *Select* colors for each land use description that most closely resembles its true landscape color.

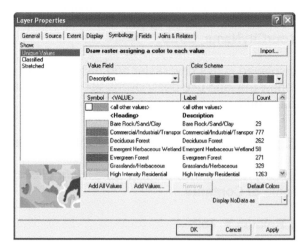

43.   When you have finished, *click* OK.

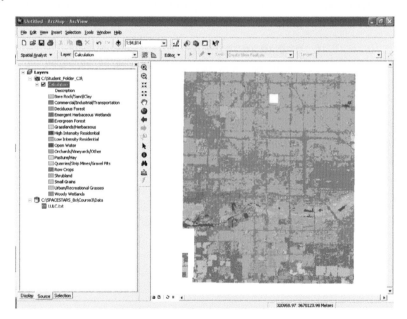

44.   *Rename* this data frame **YourCityName Land Use**.

45.   *Allow* your instructor to see your work.

46.   *Save* the project.

Analyzing Land Sensitivity

A **Land Sensitivity Analysis** is done to identify geographic conditions in an area that may pose challenges to development. For this study, we will analyze the relationship between waterbodies/wetlands, and elevation. Developments in these areas run the risk of flooding, unstable soil, and wildlife habitat destruction. By identifying those areas where these risks may occur, planners and developers can protect them from inappropriate and dangerous development.

1.   *Insert* a new **data frame** into the **S3SAL6_XX.mxd** project.

2.   *Rename* it **Land Sensitivity**.

3.   *Add* ✛ the **YourQuadName_lulc** raster grid located in your **student folder.**

4.   *Add* ✛ **YYYY_XXX_grd** raster grid (where **XXX** is your school abbreviation and **YYYY** is your county abbreviation) data layer located in the **C:\STARS\ExtToolsinSA\local** folder to the map.

5.   *Click* to view the Table of Contents in Display mode, if necessary.

To do a **Land Sensitivity Analysis**, we need to determine what elevation and land uses may pose challenges to land development. For this exercise, we will use water and wetlands as the sensitive land uses, and elevations equal to or below 20 feet *(this elevation value will vary depending on your local geography and topography).*

In case your city area extent is limited (as is the case in this example), and you want to consider land sensitivity beyond the boundaries of our own city, reset the extent to include all areas in the quad extent.

6.   ***Select Options...*** from the Spatial Analyst ▼ drop-down menu.

7.   ***Set*** **<None>** as the **Analysis mask**.

8.   ***Click*** the **Extent** tab.

9.   ***Set*** the **Analysis extent** as **Same as Display**.

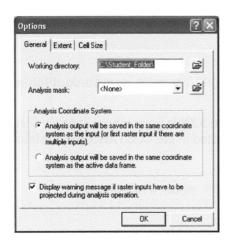 

10.   ***Click*** OK .

11.   ***Select*** **Raster Calculator...** from the Spatial Analyst ▼ drop-down menu.

In the Raster Calculator query box, you will build an equation to query the two raster grids to find those areas that meet the sensitivity criteria. To do this,

12.   ***Double-click*** **YourQuadName_LULC** to enter it into the Query box.

13.   ***Click*** the **equal sign** (=) to enter it into the Query box.

14.   ***Enter*** **11** using the keypad or the keyboard to enter it into the Query box.

15.   ***Click*** the Or button to enter it into the Query box.

16.   ***Double-click*** **YourQuadName_LULC** to enter it into the Query box.

17.   ***Click*** the **Greater than or Equal to** sign >= to enter it into the Query box.

18.   ***Enter*** **91** using the keypad or the keyboard to enter it into the Query box.

19.   ***Click*** And to enter it into the Query box.

---

20.   ***Double-click*** **YYYY_XXX_grd** to enter it into the Query box.

21.   ***Click*** the **Less than or Equal to** sign  to enter it into the Query box.

22.   ***Enter 20*** using the keypad or the keyboard to enter it into the Query box.
(This value may be different for your geographic area. Choose an appropriate elevation that may pose development challenges in your community.)

Note: For Phoenix, AZ a value of 1,000 feet was used because this area has higher elevation.

23.   When you have finished building the query, ***click*** Evaluate .

You will now have a new data layer named **Calculation** in the Table of Contents. This raster grid layer contains two data values, as follows:

False (0): This is where none of the criteria in the query has been met. In this case it includes all areas that are not either water or wetlands, and are not at or below 20 feet (or whatever elevation you used) in elevation. **Even if an area meets _one_ of the criteria set for the query, it will display as False because _both_ criteria must be met in order to result in a True reading.**

True (1): This is where all of the criteria in the query have been met. In this case it includes all areas that are either water or wetlands, **AND** are at or below 20 feet (or whatever elevation you used) in elevation.

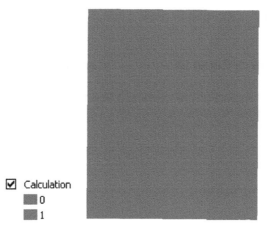

☑ Calculation
■ 0
■ 1

In this case, all of the areas in **purple** meet the land sensitivity criteria and must be excluded from any consideration for development.

In order to use this information more effectively, a shapefile mask can be created to overlay on top of any map or site plan of the area that planners and developers may use when analyzing potential sites for future development.

24.   *Select* **Convert ▶ Raster to Features** from the Spatial Analyst ▼ drop-down menu.

25.   *Set* the **Input raster:** to **Calculation**.

26.   *Set* **Value** as the **Field**.

27.   *Set* the **Output geometry type:** to **Polygon**;

28.   *Save* the **Output features:** as **YourQuadName_mask.shp** in your **student folder**.

29.   *Click* [OK]. The new shapefile is added to your map.

You will now select only those polygons which meet the sensitivity criteria, and are therefore to be excluded from any consideration for future development in the area.

30. **Select** **Select by Attributes** from the **Selection** menu.

31. **Enter** "GRIDCODE" = 1 as the selection expression.

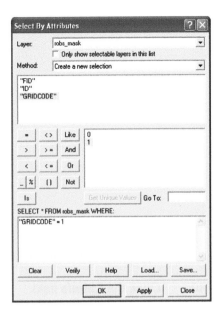

32. **Click** OK .

Only those polygons with **GRIDCODE = 1** are selected and outlined in light blue.

33. **Right click** this layer and **select** **Data ▶ Export...** to export only these **selected features** to a new shapefile **YourQuadName_mask_sel.shp**.

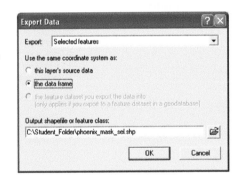

34. **Add** this new shapefile to the map when prompted.

35. **Turn off** all other layers except this new **YourQuadName_mask_sel.shp.**

36. **Change** the color of the symbol for the **YourQuadName_mask_sel** data layer to **RED** ▇.

37. To provide reference to these areas so they can be better identified on the ground, **add** ⊕ the **YYYY_XXX_str_utm##_z##.shp** street layer located in the **C:\STARS\ExtToolsinSA\local** folder.

38. **Zoom** closer to some of the **YourQuadName_mask_sel** features.

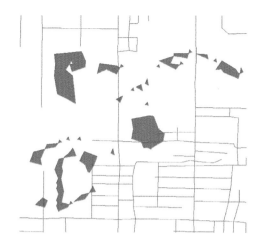

39.   **Open** **Layer Properties** for the **YYYY_XXX_str_utm##_z##** street layer.

40.   **Click** the **Labels** field.

41.   **Check** to 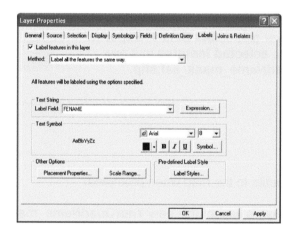 Label features in this layer . Make sure the **FENAME** field is set as the **Label Field**.

42.   **Click**   OK .

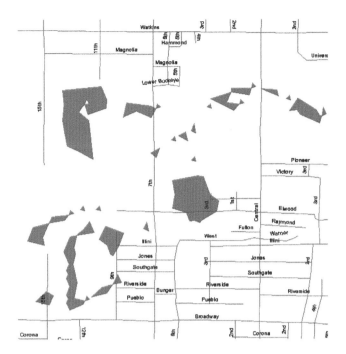

It is now possible to use this map to exclude these areas when considering future sites for development.

Creating the Map Layout:

1. *Switch* to **Layout View**.

2. *Change* the **Page Orientation** to **Landscape**.

3. *Move* the data frames to appropriate places on the layout page.

4. *Insert* the following **title** on the map layout in **24-point bold** type:

   Spatial Analyst Mini Project – Environmental Issues.

4. *Insert* a **legend** on the map layout page for each of the data frames.

5. *Insert* a **scale bar <u>for each of the data frames</u>** on the layout page using display units that are appropriate for the map layout.

6. *Insert* a **north arrow** on the layout page.

7. *Place* your **name** and the **date** on the layout page.

8. *Insert* a **neatline** as a border on the map layout page.

9.   ***Export*** the map as **S3SAL6_XX** (where **XX** is your initials) in **JPEG** format to your **student folder**.

10.   ***Print*** the layout.

11.   ***Save*** 🖫 the project.

12.   Allow your instructor to see your work before you exit ArcMap.

13.   ***Exit*** ArcMap.

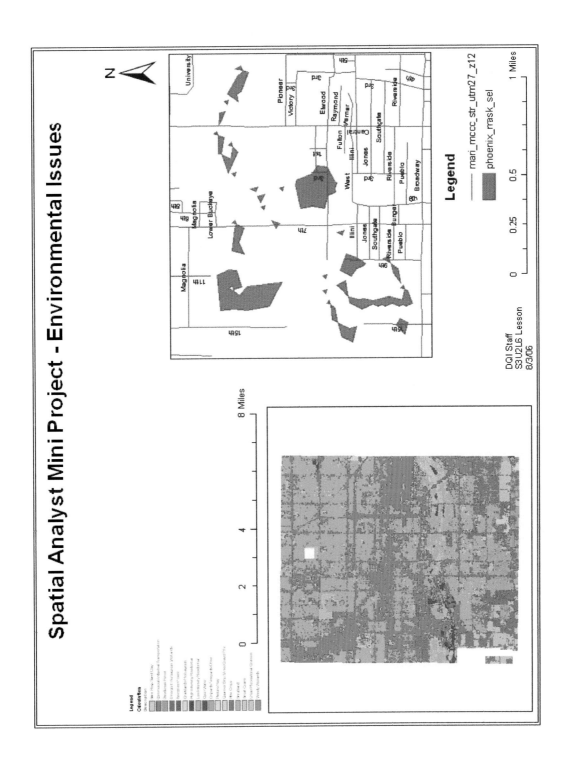

Lesson 6: Applications - Spatial Analyst Project Lesson Review

Key Terms
Use the lesson or index provided in the back of the book to define each of the following terms.

      1. Land Sensitivity Analysis

Global Concepts
Use the information from the lesson to answer the following questions. You may need to answer these on the back of this page or on your own paper.

    2.   For your city, what land use type was the most common and why do you think that is?

    3.   Is waterbodies, wetlands and elevation a severe issue for your city? If so, please explain why, if not, please suggest other issues that might be more problematic.

Let's Talk About It...
Answer the following question and share the responses with your instructor and classmates.

    4.   If you were the planner for your town, what kind of analysis would you most want to do and why?

Glossary

Glossary

Aspect: Downslope direction of landscape

Base Contour: The value from which to begin generating contours

Cell Statistics: Calculations made about each cell on a raster grid among multiple rasters in the same geographic area

Contour Interval: Distance between contour lines;the wider the range in elevation, the higher the contour interval

Contours: Polylines that connect points of equal value

Cost-Weighted Distance: Distance measured taking a cost(such as topology) into consideration, each cell is the least accumulated cost of traveling from each cell to the source

Cut/Fill Analysis: Shows where there are losses and gains in surface area between DEMs of a location at two different points in time

Direction Grid: Data set showing the direction each part of the grid is from the points of interest

Euclidean Distance: Another term for straight-line distance

Hillshade:	View of a surface as if there is a light source above it, simulating the relief of a surface under sunlight
Interpolation:	The process of taking known data point values and estimating the values across an entire surface
Inverse Distance Weighted:	Estimating the value of unknown cells using the distance and values of nearby known cells; each cell is inversely weighted in proportion to the distance from other cells
Kernel Density	Sum of all points in the search area divided by the search area?s size, but the cells closest to the point are weighted more heavily than those further away
Kreiging:	Assumes there is a spatial correlation between the distance and directions of the data points. A formula is used to explore that relationship and create the surface.
Land Sensitivity Analysis:	Analysis done to identify geographic conditions in an area that may pose challenges to development
Neighborhood Statistics:	Calculations made about groups of cells(a neighborhood) in a raster grid
Point Density:	Synonymous with simple density
Simple Density:	Sum of all points in the search area divided by the search area's size

Slope:	Rate of elevation change;
Spline:	Using a mathematical formula to smooth a curve by "connecting the dots" of known sample points
Straight Line Distance:	Distance measured "as the crow flies", without consideration of any potential costs, including topology
Viewshed:	Places in a study area that can be seen from a particular observation point
Zonal Statistics:	The statistical analysis of the relationship between two different features